# Blockchain-based Internet of Things

This book presents an overview of the blockchain-based Internet of Things systems, along with the opportunities, challenges, and solutions in diverse fields such as business, education, agriculture, and healthcare. It discusses scalability, security, layers, threats, and countermeasures in blockchain-based Internet of Things network.

**Features:**

- Elaborates on the opportunities presented by combining blockchain with artificial intelligence on the Internet of Things systems in the management of food systems and drug supply chains.
- Explains the management of computationally intensive tasks in blockchain-based Internet of Things through the development of lightweight protocols.
- Presents various applications in fields including logistics and the supply chain, automobile industry, smart housing, shared economy, and agriculture.
- Provides insights into blockchain-based Internet of Things systems, along with their features, vulnerabilities, and architectural flaws.

The text is primarily written for graduate students, and academic researchers working in the fields of computer science and engineering, electrical engineering, and information technology.

I0473121

# Blockchain-based Internet of Things

# Opportunities, Challenges and Solutions

Edited by
Iraq Ahmad Reshi and Sahil Sholla

## CRC Press

Taylor & Francis Group
Boca Raton London New York

CRC Press is an imprint of the
Taylor & Francis Group, an informa business

A CHAPMAN & HALL BOOK

Front cover image: PATTYARIYA/Shutterstock

First edition published 2024
by CRC Press
2385 NW Executive Center Drive, Suite 320, Boca Raton FL 33431

and by CRC Press
4 Park Square, Milton Park, Abingdon, Oxon, OX14 4RN

*CRC Press is an imprint of Taylor & Francis Group, LLC*

ISBN: 978-1-032-52487-0 (hbk)
ISBN: 978-1-032-52538-9 (pbk)
ISBN: 978-1-003-40709-6 (ebk)

DOI: 10.1201/9781003407096

Typeset in Times
by Newgen Publishing UK

# Contents

# Preface

In an era defined by groundbreaking technological advancements, the fusion of blockchain and the Internet of Things (IoT) has emerged as a dynamic and transformative force. The rapid proliferation of smart devices, coupled with the decentralized nature of blockchain technology, has ushered in a new paradigm that promises to revolutionize the way we interact with and perceive the digital realm. The convergence of blockchain and IoT has given rise to a landscape of unprecedented impact, growth, and potential. As smart devices seamlessly interconnect various facets of our lives—spanning from homes and workplaces to vehicles and industries—a world of endless possibilities unfolds before us. However, amid this remarkable progress, a profound challenge looms on the horizon. This book dispels the misconception that mere backups can safeguard individuals and organizations from the ever-evolving threat of cyberattacks. By delving deep into the intricate fabric of the blockchain-based IoT, it exposes the critical security flaws inherent in this intricate ecosystem. From the fundamental physical layer to the intricate application level, every facet of security is rigorously examined. Navigating the complex terrain of blockchain-based IoT security requires a comprehensive approach. This book elucidates the multifaceted nature of this challenge, investigating diverse security and privacy threats that cast a shadow over this innovative landscape. It illuminates the path toward designing robust, resilient solutions that stand as bulwarks against potential breaches. In this exploration of the future, hands-on insights into fundamental concepts of various open-source tools in the realm of blockchain-based IoT security take center stage. Readers are invited to embark on a journey of discovery, equipping themselves with the knowledge to forge ahead in a realm where pioneering research directions intersect with tangible opportunities.

The pages of this book offer a rich tapestry of topics that traverse the intricate terrain of the blockchain-based IoT. It serves as both a compass for those exploring captivating research avenues and a guiding light toward uncharted directions awaiting further investigation. As we peer into the convergence of blockchain and IoT, we are beckoned to acknowledge the challenges, grasp the research frontiers, and seize the boundless opportunities that lie ahead. This book is a roadmap for the curious, the innovative, and the intrepid—a testament to the unyielding human spirit that propels us toward a future where security, connectivity, and possibility intertwine in unprecedented harmony. Following are the topics covered in this book.

- Blockchain: Concept and Emergence
- Blockchains in IoT: Introduction, Features, and Vulnerabilities
- Blockchain-Based Internet of Things (B-IoT): Challenges, Solutions, Opportunities, Open Research Questions, and Future Trends
- Revolutionizing IoT with Blockchain: A State-of-the-Art Review
- A Framework for Smart and Resilient Supply Chains Based on Blockchain and the Internet of Things
- Pharma-Blocks: Blockchain-IoT Platform for Pharmaceutical Sector

- Edge Intelligence Decentralized Blockchain-Based Internet of Things (B-IoT) for Sustainable Healthcare
- Analysis of AI Embedded Block Chain Security Model for Healthcare and Financial Transactions
- Implementation of a Blockchain-Based Secure Cloud Computing Mechanism for Transactions
- Geolocation-Based Smart Land Registry Process with Privacy Preservation Using Blockchain Technology and IPFS
- Safeguarding Digital Environments: Harnessing the Power of Blockchain for Enhanced Malware Detection and IoT Security
- Synergizing Information Diffusion: Exploring IoT and Blockchain Integration in Online Social Networks

# Acknowledgments

I offer my heartfelt gratitude to the Almighty Allah for granting me the strength, guidance, and perseverance throughout this journey to edit this book.

To my beloved parents, Ammi and Abuji, your unwavering love, encouragement, prayers, and sacrifices have been my source of inspiration. I am forever indebted to you for your endless support.

I extend my sincere appreciation to my supervisor, co-author, and mentor, Dr. Sahil Sholla, for their invaluable guidance, mentorship, and insightful contributions that have shaped this work.

To my dear sister, my life partner, our little Huzaif, thank you for your patience, understanding, and constant motivation. Your presence has been my source of joy and encouragement.

Special thanks to my colleagues, friends, faculty members, and lab mates at our department of CSE, Islamic University of Science and Technology, whose camaraderie, discussions, and support have been invaluable.

Finally, heartfelt appreciation to all the contributors who, directly or indirectly, aided in the completion of this book. Your assistance is genuinely appreciated and valued.

Thank you all for being an integral part of this endeavor.

# Contributors

**Md Ehsan Asgar**
Department of Mechanical Engineering
HMR Institute of Technology and
  Management
Hamidpur, New Delhi, India

**Asif Ali Banka**
Department of CSE
Islamic University of Science &
  Technology
Awantipora, Kashmir, India

**Mahmonir Bayanati**
Technology and Industrial Management
Health and Industry Research Center
West Tehran Branch, Islamic Azad
  University
Tehran, Iran

**Rahul Bhandari**
Schaefer School of Engineering and
  Science
Stevens Institute of Technology
New Jersey, United States

**Nowsheena Bhat**
Department of Computer Sciences
University of Kashmir
J&K, India

**Preeti Chandrakar**
Department of Computer Science &
  Engineering
National Institute of Technology, Raipur
Chhattisgarh, India

**Narendra Kumar Dewangan**
Department of Computer Science &
  Engineering
National Institute of Technology, Raipur
Chhattisgarh, India

**Aaquib Hussain Ganai**
Department of Computer Sciences
University of Kashmir
J&K, India

**Rana Hashmy**
Department of Computer Sciences
University of Kashmir
Srinagar, India

**Javid Ghahremani-Nahr**
Academic Center for Education
Culture and Research (ACECR)
Tabriz, Iran

**Neeraj Gupta**
Panipat Institute of Engineering &
  Technology
Panipat, Haryana, India

**Muhammad Zulkifl Hasan**
Faculty of Computer Science and
  Information Technology
University of Central Punjab
Lahore, Pakistan

**Muhammad Zunnurain Hussain**
Department of Computer Science
Bahria University Lahore Campus
Pakistan

**Aatifa Jan**
Department of Information Technology
Central University of Kashmir
Ganderbal, Kashmir, India

**Hilal Ahmad Khanday**
Department of Computer Sciences
University of Kashmir
J&K, India

**Jeya Mala**
School of Computer Science and
  Engineering
Vellore Institute of Technology
  (Deemed University)
Chennai, Tamil Nadu, India

**Adil Mudasir Malla**
Department of CSE
Islamic University of Science &
  Technology
Awantipora, Kashmir, India

**Seshadri Mohan**
Systems Engineering Department
University of Arkansas at Little Rock
Little Rock, Arkansas, United States

**Mudasir Mohd**
Department of Computer Sciences
University of Kashmir
J&K, India

**Ghulam Mustafa**
Department of Computer Science
University of Central Punjab
Lahore, Pakistan

**Zahoor Ahmad Najar**
Department of Information technology
Central University of Kashmir
Ganderbal, Kashmir, India

**Hamed Nozari**
Department of Management
Azad University of Dubai
Dubai, UAE

**Adnan Nabeel Qureshi**
Department of Computer Science
Birmingham City University
United Kingdom

**Maryam Rahmaty**
Department of Management
Chalous Branch, Islamic Azad
  University
Chalous, Iran

**A. Pradeep Reynold**
Department of Safety
ASET College of Science and
  Technology
Chennai, Tamil Nadu, India

**Sania**
Department of Computer Science and
  Engineering
The NorthCap University
Haryana, India

**Wasswa Shafik**
Universiti Brunei Darussalam, Gadong,
  Brunei Darussalam
Dig Connectivity Research Laboratory
  (DCRLab), Kampala, Uganda

**Preeti Sharma**
Department of Computer Science and
  Engineering
The NorthCap University
Haryana, India

**Sachin Sharma**
Department of Computer Science and
  Engineering
Graphic Era (Deemed to be University)
Uttarakhand, India

**Ranu Tyagi**
Department of Computer Science and
  Engineering
Graphic Era (Deemed to be University)
Uttarakhand, India

**Mohsin Altaf Wani**
Department of Computer
    Sciences
University of Kashmir
J&K, India

**Saba Zaidi**
Department of Computer Science and
    Engineering
Panipat Institute of Engineering and
    Technology
Panipat, Haryana, India

# About the Editors

**Iraq Ahmad Reshi** is Research Scholar at the Department of Computer Science and Engineering, Islamic University of Science and Technology, Kashmir, India. He has pursued his BTech from NIT Srinagar, India, and MTech from Central University of Kashmir, Kashmir, India. His research interests include blockchain, artificial intelligence, and Internet of Things.

**Sahil Sholla** is Assistant Professor at the Department of Computer Science and Engineering, Islamic University of Science and Technology, Kashmir, India. He has received his PhD from NIT Srinagar, India. His research focuses on technology ethics, security, blockchain, artificial intelligence, and Internet of Things.

# 1 Blockchain
## *Concept and Emergence*

Aatifa Jan and Zahoor Ahmad Najar

Department of Information Technology, Central University of Kashmir, Kashmir, India

## 1.1 WHAT IS BLOCKCHAIN?

The term "blockchain technology" was first described in 1991 by Haber and Stornetta just as "a cryptographically protected sequence of blocks" (Saxena et al. 2021). Nevertheless, Satoshi Nakamoto, author of the paper "Bitcoin: A peer-to-peer electronic cash system," is credited for popularizing blockchain technology (Nakamoto 2008). Nakamoto created the "creation block" and the Bitcoin social network in 2009 (Wang et al. 2019). The term "blockchain" comprises two terms, "block" and "chain." Thus, it is comprehensibly the chain of blocks. A block is a data structure where the number of transactions is stored. A Cryptographic hash of a preceding block is tied to a succeeding block, forming the chain of blocks (Wüst and Gervais 2018). So, blockchain is a shared, peer-to-peer, unalterable, moreover generally public, encrypted digital ledger system ("Blockchain"; Singh and Tripathi 2019).

Blockchain network members take care of the peer-to-peer decentralized database and do not need to be maintained by a central authority or a trusted third party. The technology ensures reliability by making tampering data stored in the block visible (Singh and Tripathi 2019). In simple words, architecturally, blockchain can be considered as the distributed database spread among the members of the decentralized network, including timestamped blocks of transactions linked together in a train of blocks by pointing to the block before them. Operatively, they record independently available data transparently and tamperproof while still providing transactional service (Gorbatyuk and Gils 2022). Figure 1.1 shows the structure of a block in a blockchain.

Description of technical words with their definitions is presented in Table 1.1.

## 1.2 CLASSIFICATION OF BLOCKCHAIN TECHNOLOGY

Several kinds of blockchains house various node types. While some nodes can play a more active role (write-/commit-permission), others are more passive (read-permission). A node with "read" permission can access the database and view transactions, but a node with "write" permission can generate transactions and transmit them over the blockchain network. A node with the "commit" permission

DOI: 10.1201/9781003407096-1

**TABLE 1.1**
**Technical Terms**

| Term | Meaning |
| --- | --- |
| Node | The Blockchain system's database. |
| Smart contracts | Digital contracts enforced by themselves or breaching the agreement are unreasonably expensive (Wüst and Gervais 2018). |
| Timestamp | Utilizing the time and date stored in the computer system, an electronic time stamp for the transaction is generated. |
| Hash | It is the output of the One-way hash function to verify the legitimacy of a transaction or data. |

can modify the ledger, such as a miner or a validator node, as shown in Table 1.2 (Gorbatyuk and Gils 2022).

## 1.3  CHARACTERISTICS OF THE BLOCKCHAIN TECHNOLOGY

A.  Immutability

The meaning of immutability is permanence, something which is unmodifiable. It is the utmost-yearned feature that upholds the resilience of blockchain technology. When the transactions are put onto the blockchain, nobody can delete or update them afterward. Thus, it renders blockchain technology tamperproof (Singhal et al. 2018). To nullify an erroneous transaction, one can add a fresh transaction to cancel the fault in the previous transaction; however, both old and new transactions are seeable to every network participant. As transaction gets stored on the chain, nigh everybody has a clone of it. When the chain grows, and more fresh blocks are added to the blockchain, immutability gets mounted more and more, and eventually, it grows into an entirely immutable ledger. For some lousy actor to mischievously alter a block, it becomes approximately exhaustive if one block is altered, and then all the blocks after it need to be modified (Singhal et al. 2018).

B.  Decentralization

The meaning of decentralization in blockchain technology is that no centric or intermediate power(control) governs the system; rather, the decentralized blockchain networks are controlled by rules and participants themselves. Blockchain systems are autonomous (Singhal et al. 2018). There is no participant mightier and more authoritative than other participants, as shown in Figure 1.2. Each network member possesses similar privileges in the system, and decisions and rulings are implemented when most participants arrive at a mutual consensus (Singhal et al. 2018).

C.  Distributed

Distributed blockchain technology means that all members of a blockchain network own a replica of the database, making the blockchain network a fully see-through and transparent system. When changes happen in this distributed

**TABLE 1.2**
**Classification and Attributes of Blockchain Technology**

| Type of Blockchain | Permissionless | | Permissioned | |
|---|---|---|---|---|
| **Public** | PUBLIC- PERMISSIONLESS Participation (read- permission) is uncontrolled while (write and commit- permissions) are bestowed to every Node. | ATTRIBUTES<br>• Inferior Extensibility.<br>• Completely Decentralized.<br>• Sluggish Transaction rapidity.<br>• Dropped Productivity. | PUBLIC-PERMISSIONED Participation (read-permission) is uncontrolled while as (write and commit-permissions) are bestowed to either one node or a restricted count of nodes. | ATTRIBUTES<br>• Lifted Extensibility.<br>• Guardingly Decentralized.<br>• Upgraded Transaction rapidity.<br>• Fine Productivity. |
| **Private** | PRIVATE- PERMISSIONLESS Participation (read-permission) is controlled while as (write and commit-permissions) are bestowed to every approved node. | ATTRIBUTES<br>• Fine Extensibility.<br>• Judiciously Decentralized.<br>• Good Transaction rapidity.<br>• Increased Productivity. | PRIVATE- PERMISSIONED: Participation (read-permission) is controlled, while (write and commit-permissions) are bestowed to either one node or a restricted count of nodes. | ATTRIBUTES<br>• Superior Extensibility.<br>• Partly Decentralized.<br>• Elevated Transaction Rapidity.<br>• Proficient Productivity |

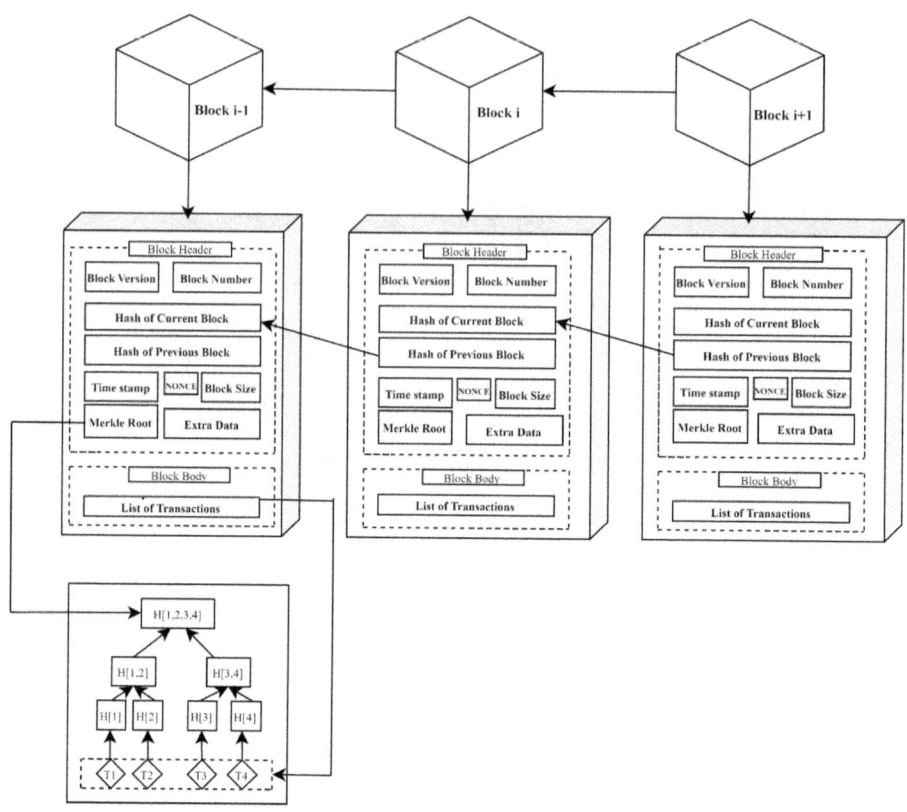

**FIGURE 1.1**   Block structure.

blockchain technology, they spread speedily, therefore allowing blockchain networks to be alteration-safe (GeeksforGeeks 2020). Every node that participates possesses the same clone of the distributed database, so it accrues in a system where data will remain perpetual even when a node stops working (Gorbatyuk and Gils 2022).

D.  Consensus

Blockchain technology is booming by consensus algorithms. The meaning of consensus is common consent; in blockchain networks, the consensus is there to build decisions that are not only speedy but also not prejudiced (GeeksforGeeks 2020; 101 Blockchains n.d.). The consensus algorithm governs, manages, and directs the decentralized blockchain network without a central authority. The consensus algorithm is one of the reasons behind the blockchain networks existing untrustworthy. There is no need for the network members to trust one another, but they can have faith in the consensus algorithm. When numerous nodes check the transaction's legitimacy, the consensus makes the process seamless (101 Blockchains n.d.). Although there

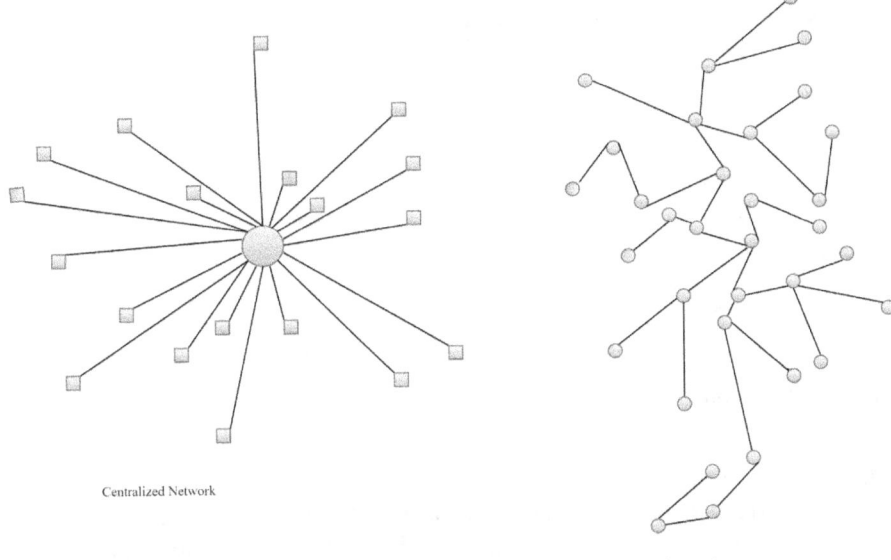

Centralized Network

Decentralized Network

**FIGURE 1.2** Centralized and decentralized network.

are various consensus algorithms, each differs from another while decision-making; all blockchain networks must have one to run the system efficiently (GeeksforGeeks 2020; 101 Blockchains n.d.).

E.  Transparency
The blockchain ledger is open, making it transparent. Any authorized member of the network can read and examine the transactions in the blockchain network (GeeksforGeeks 2020), rendering the blockchain system tamper-evident, which does not support malicious modifications and is utmost where nobody gets any perks from the network (101 Blockchains n.d.).

F.  Auditable
As in a blockchain network, the blocks are chained together owing to the employment of hash values. The blockchain's beginning block, which has no previous hash, is called the genesis block, and new blocks get added to this genesis block. So, it not only makes tracking the transaction quite easy but also validates the incorruptibility of the transaction (Gorbatyuk and Gils 2022; Singhal et al. 2018).

## 1.4  WORKING OF BLOCKCHAIN TECHNOLOGY

The subsequent steps summarize the sophisticated fundamental blockchain technology execution, as depicted in Figure 1.3.

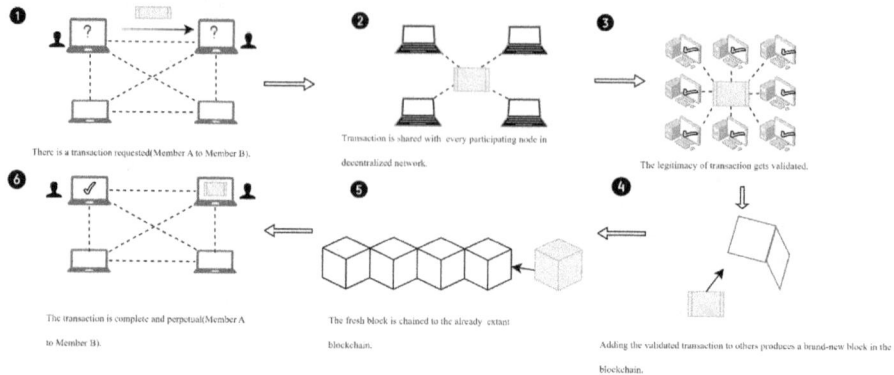

There is a transaction requested( Member A to Member B).

Transaction is shared with every participating node in decentralized network.

The legitimacy of transaction gets validated.

The transaction is complete and perpetual(Member A to Member B).

The fresh block is chained to the already extant blockchain.

Adding the validated transaction to others produces a brand-new block in the blockchain.

**FIGURE 1.3** Working of the blockchain.

**Step 1**: Storing of the transaction.
The distributed ledger network's nodes establish and transmit transactions. They shift the distributed ledger's status from one phase to another (Wüst and Gervais 2018). They exhibit the migration of offline or online assets from one participant to the other participant in the network. They get to be stored in the blocks as data, and it depends upon the creator of the blockchain on what data it can store (Griswold 2021). It can store details about the participants of the transaction, at what time did transaction occur, the number of assets that got transacted, some more details about the preconditions, etc.

**Step 2:** Acquiring consensus.
Every transaction is protected using cryptography and timestamping and approved by all the blockchain network members employing the consensus algorithms (list of rulings) (Sadowsky 2021). The largest number of blockchain network members should agree on the transaction's legitimacy. Any transaction not considered genuine by all the authorized participants of the network will not be eventually recorded onto the blockchain. Owing to its immutability, once a transaction is stored in the database, it can neither be removed nor updated afterward.

**Step 3**: Connecting of blocks.
When someone who participates in the distributed ledger network arrives at a consensus about the genuineness of the transaction in question, that particular transaction is written onto the blockchain blocks. The list of transactions and the cryptographic hash value are also added to the data block. This hash value or message digest is the backbone behind creating a chain of blocks in chronological order. When data in the block gets tampered with, it becomes visible as the hash does not remain the same as the previous one. Thus, each hash value is used to identify its corresponding block. When a new block joins the existing blockchain network, it not only proves its authenticity but also of the previous blocks and eventually of, the whole chain.

**Step 4**: Distribute the ledger.

The most recent replica of the blockchain is disseminated to entire participants of the decentralized blockchain network.

## 1.5 APPLICATION OF THE BLOCKCHAIN TECHNOLOGY

According to claims and perceptions, the science and technology referred to as blockchain is one of the revolutionary technologies that will significantly influence our lives in the years and decades to come. Blockchain technology can revolutionize most sectors in several ways with positive effects.

A. Asset management.

Asset administration appears similar to the economy, where a decentralized ledger bears a significant part. The endless possibilities of the digital ledger in asset management are altogether to the next level. Managing assets involves dealing with assets and trading the various types of assets that one can possess, such as goods, bond funds, property, investments, etc. Blockchain can be a real aid by employing disintermediation, especially when the different intermediaries (middlemen) get involved in the various asset management processes, making the whole process see-through. The distributed digital ledger provides a simple, comprehensive process that leaves no space for mistakes (GeeksforGeeks 2020).

B. Healthcare.

The patient is at the center of the medical system. The patient should also get truthful information on their health state and any procedures done. An accurate patient medical history can mean the difference between life and death. Data security and privacy for healthcare are essential. Prescription medication serial and batch numbers may be tracked using blockchain technology. A digital ID or a numerical key will be provided to the patient to access these documents. Blockchain enables patients to decide who may access their data and how it is used. It is also possible to preserve the patient's diagnosis to access their medical history (Knowledgenile 2020). Blockchain will not only help with the record-keeping of patients' data. However, with its incredible features, blockchain can help forge a ground that guarantees trustworthy record-keeping. As digital healthcare reports get dispersed during transmission among various healthcare institutes, it has given the biggest headache to healthcare (Chen et al. 2018). Forbes (2018) says that blockchain with artificial intelligence could boon the healthcare industry by presenting solutions to various problems.

C. Cryptocurrency.

Especially in the realm of cryptocurrencies, the financial industry is one of the most active uses of the blockchain. Numerous cryptocurrencies have appeared since the first one, bitcoin, entered the blockchain (Nakamoto 2008). Blockchain's first application was the cryptocurrency, namely, Bitcoin. Cryptocurrencies have the upper hand over fiat currency as it has no geographic constraint. Therefore, cryptocurrency can be used for trade-in from and around the world (GeeksforGeeks 2020). The present thriving cryptocurrency industry is made up of several other cryptocurrencies that came into being that have arisen and have better functionality. Among them, Ethereum (Ethereum 2013) became highly recognizable as it came out as a platform supporting smart contract building in 2015. With the advent of smart contracts, blockchain technology found its utilizations in the sharing economy and the Internet of Things (IoT). The digital ledger is increasingly employed in financial services and cryptocurrencies. According to the Australian securities exchange, the clearing system would be replaced with bitcoin technology to lower transaction costs and speed up and secure transactions. The London-based trading business Oxygen has announced the opening of its repos blockchain platform (Chen et al. 2018). Making cross-border payments may be lengthy and convoluted, and the money may take several days to reach its destination. The elimination of intermediaries through person-to-person transfers made possible by the distributed ledger has helped to streamline such international payments. The distributed ledger offerings from several exchange companies may be utilized to move payments worldwide in a single day (GeeksforGeeks 2020).

D. Supply chain management.

Supply chain management (SCM) controls the movement of goods needed to generate a specific good, including all stages of intermediate production and storage up to delivery to the point of consumption. Several businesses engage in worldwide commerce and interaction in a typical supply chain. Because of this complexity, controlling the inventories, procedures, and failure detection comes at a high cost (Wüst and Gervais 2018). Therefore, blockchain may make the supply chain more effective overall. The spot of things along the supply chain may be precisely known due to blockchain technology, eliminating the requirement for paper-based trials. The blockchain's paperless method aids in preventing data loss and damage. Blockchain is also useful for gauging product quality while being produced (Knowledgenile 2020).

E. Internet of Things.

IoT can be defined as the network of gadgets having a network connection that enables communication and information gathering, which can be used to guide choices. Any set of "things" networked together form a system for the IoT. The distributed ledger is necessary to ensure the safety of this massively connected system. In such a scenario, blockchain technology can ensure that the information acquired by IoT gadgets remains confidential and is only exposed to those with the proper permissions for viewing and using the information in question (GeeksforGeeks 2020).

F. Proof of intellectual property ownership.

Blockchain application cases for proving intellectual property ownership are frequently suggested and simple. The author of an online thing can use a distributed ledger that serves as a sort of time tagging tool and subsequently authenticate ownership by agreeing to the electronic item jointly with their recognition, for example, via a hash value and broadcasting the agreement onto the distributed ledger. Thus, it is possible to demonstrate the object's existence and connection to the specific identity. However, a public blockchain makes it easier to provide distributed verification without revealing specifics of the underlying item (Wüst and Gervais 2018). Royalty payments and copyrights remain significant issues in the artistic fields, including soundtracks, movies, etc. These are simply demonstrations of ingenuity, and it does not appear like they've got much dealing with the decentralized ledger. Such technology is vital for assuring access and protection in the artistic sector. Whenever the producers of the created pieces are not correctly acknowledged, there have been numerous instances of piracy in songs, the movie industry, art, etc. The distributed ledger can resolve the issue by keeping an exhaustive record of artist rights. In addition to being translucent, it can provide a secure archive of artist revenues and contracts with important production businesses. You might also manage payments for royalty fees using electronic money such as Bitcoin (GeeksforGeeks 2020).

## 1.6   WHAT IS THE INTERNET OF THINGS?

The Internet is the network of networks that links people to information. In contrast, the IoT is an interconnected system of tangible items—"things" which are incorporated with software, detectors, and sensing, along with additional technology in order to get connected to various other gadgets and systems to share data with them across the Internet (Hussein 2019; Oracle n.d.). (Hussein 2019) says that the IoT is a network of uniquely addressable physical objects with a range of sensing, processing, and actuating abilities that may cooperate and communicate using the Internet as their common platform. Such devices include anything from common domestic items to high-tech industrial equipment. Therefore, the primary goal of the IoT is to enable connections between things and people at any time and everywhere, utilizing any network or service. The IoT has slowly begun to be viewed as the next step in the Internet evolution. IoT might make it easy for ordinary electronics to be linked to the Internet to accomplish various missions—thus generating a prospectus for an automatic and more effective system.

## 1.7   WHAT MAKES THE IOT OF SUCH IMPORTANCE?

The IoT is among several important modern technologies that have recently appeared. Because it is now possible to connect commonplace goods to the web via integrated equipment, such as residential appliances, cars, heating and cooling systems, etc., continuous connectivity among people, processes, and elements is now conceivable (Oracle n.d.). The IoT has a significant impact on both business and daily life. It enables machines to do more strenuous labour, takes over boring and energy-sucking chores, and improves the wellness, efficiency, and convenience of living. IoT acts as a network that includes multiple "linked" gadgets, as illustrated in Figure 1.4. In contrast, the Internet has grown into connecting different devices rather than simply connecting Pcs. Today, an extensive range of equipment, notably smartphones, autos,

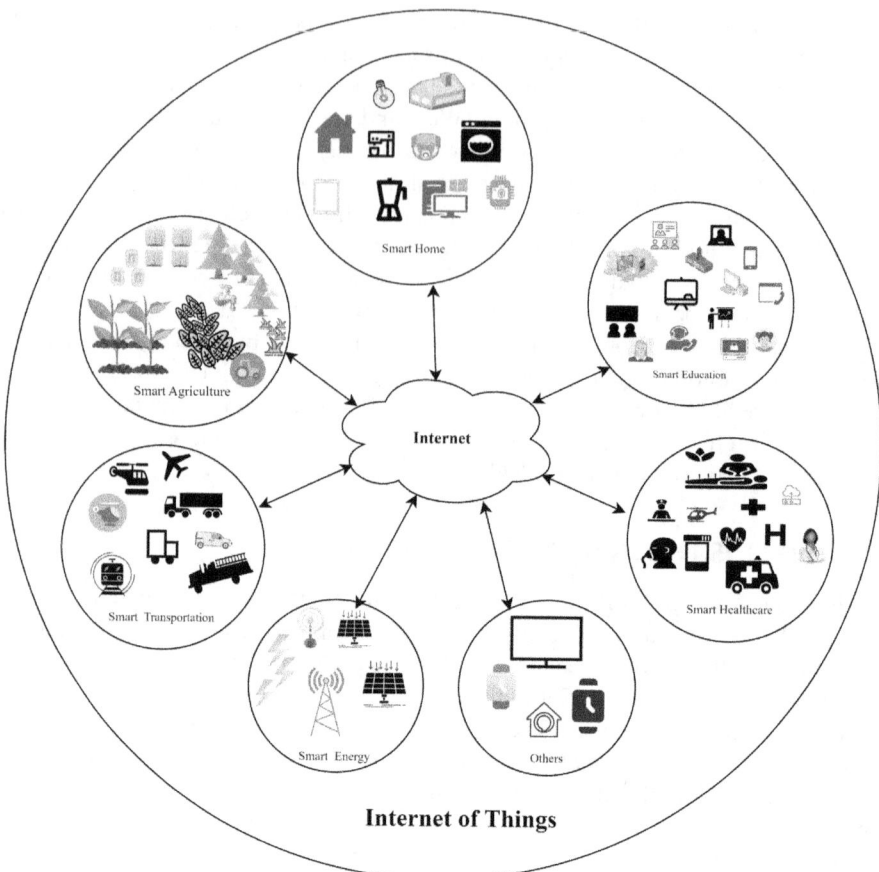

Smart Home

Smart Education

Smart Agriculture

Internet

Smart Healthcare

Smart Transportation

Smart Energy

Others

**Internet of Things**

**FIGURE 1.4**   A system of networks can be used to describe the Internet of Things.

manufacturing equipment, recording devices, items to play with, structures, kitchen appliances, and countless more, all can share material online. No matter their shapes or intended uses, these devices are capable of administration, directing, guiding, watching in real-time, and process supervision. People-to-people linkages are a minor aspect of a bigger movement approaching an amalgamation of the tangible and digital realms (Hussein 2019).

## 1.8   COMPONENTS OF IOT

As outlined below, an IoT system comprises four main parts: Detectors, connectivity, the handling of information, and interface for users.

A.   Detectors
     Basic components of an IoT include sensors or devices that gather information from objects with an IP address. These gadgets might be as straightforward

as installed climate and moisture monitors in structures or as sophisticated as smart automobiles. These parts mostly gather information from the associated surroundings, such as climate. These sensors and gadgets operate as a unit rather than individually. Devices that cluster data from exterior settings and then communicate it to the IoT ecosystem make up an IoT structure. Sensors can be used in various settings, including medical (for instance, to keep track of patient's health conditions), etc. (Al_Barazanchi et al. 2022).

B. Connectivity

Connection is a crucial part of the architecture of the IoT. Items from the IoT must be linked to the IoT infrastructure. It is essential to remain assured that anyone, anywhere, at any time, can do this. Examples include interactions between people using devices connected to the Internet, like cell phones, additional gadgets, etc., and communications made among particular Internet-connected entities like routers and gateways, gauges, etc. ("Characteristics of Internet of Things").

C. Handling of information:

Data is analyzed using software and afterward deposited in the cloud. This gathered data must be analyzed, scrubbed, and examined in a customized process before use. Myriad data is directed via each edge gateway preceding processing to prevent the entire system from being swamped (Al_Barazanchi et al. 2022). The IoT generates real-time data in a variety of formats. Large volumes of IoT-generated information must be analyzed, examined, and categorized beforehand to be utilized in making choices. Therefore, several methods must be established for locating and transforming unprocessed data into forms that can be utilized. Artificial intelligence and machine learning are two of these methods, both of which have been applied (Shang et al. 2020, 440–444).

D. User interface (UI):

After being organized and sanitized, the gathered data should also be employed to notify users. A UI provides this function by enabling end users to assess the information gathered dynamically. Because of this, these end users could also respond to system inputs according to applications developed for the IoT. An end user could respond with input, for instance, by modifying specific parameters (such as "controlling" or "regulating" the clime in the cold wallet) after getting a notification on their phone. Certain tasks can be automatic in some situations rather than requiring user input. For instance, it is possible to control the temperature in cold storage based on established criteria rather than patiently awaiting user input (Al_Barazanchi et al. 2022).

## 1.9 COMPLICATIONS SURROUNDING SECURITY IN THE INTERNET OF THINGS

Security has always been regarded as a concern with the IoT. The top worries regarding safety with IoT networks are included in the list below:

A.  Data quantity
    Many applications for the IoT, like smart electricity systems and metropol-
    itan areas, generate enormous amounts of sensitive data, making them an
    attractive spot for escalating security dangers.
B.  Accendibility
    Typically, a significant number of bodies play a role in the operation of an IoT
    system. As a result, network-wide privacy and safety defense mechanisms
    must scale well.
C.  Controlled independently
    Traditional information systems demand user settings. Nevertheless,
    the configurations in the IoT systems' endpoints must be established
    independently.
D.  Attack-withstanding
    The endpoints in an IoT system are often tiny and have minimal or no struc-
    tural protection. For instance, natural catastrophes might rob or damage cell
    phones, tiny sensor gadgets, or stationary equipment.
E.  The resources are restricted
    These gadgets can't even easily enable the execution of common safety
    mechanisms like AES or other cutting-edge privacy-safeguarding techniques
    because of their limited computing power and storage capacity.
F.  Confidentiality safeguarding
    Nodes in the IoT contain important information that must be kept private and
    shielded from identification, linking, and tracking (Saxena et al. 2021). The
    primary worry in the current linked world is safeguarding privacy due to the
    constant processing, transmission, collection, and exploitation of information
    by large businesses employing a variety of IoT gadgets.
G.  Interoperability
    The creation and use of safety protocols in IoT networks shouldn't entirely
    limit the functionalities of nodes within them. Incompatible gadgets cannot be
    used in diverse environments, which makes it impossible to construct extend-
    domain applications for the IoT, and users are left unhappy. These are only
    a few serious technical problems that could arise from an interoperability
    problem.
H.  Heterogeneity
    The IoT links various identities with varying levels of sophistication and cap-
    acity. Additionally, those gadgets vary in functional requirements, techno-
    logical interfaces, and release variants. To link diverse networks and objects,
    protocols for the IoT have to manage various objects in various settings.
I.  Anonymity
    Despite the openness of blockchain transactions, the network can preserve
    some integrity by providing participants with anonymized addresses. Due to
    the traceability of these addresses, blockchain-based systems can guarantee
    secrecy to an extent. Blockchain can, therefore, only preserve anonymity, not
    complete secrecy (Saxena et al. 2021).

## 1.10 THE NECESSITY FOR INTERNET OF THINGS AND BLOCKCHAIN TECHNOLOGY CONVERGENCE

Although the IoT has a wide range of uses, encompassing healthcare, automobile traffic control, smart houses, smart cities, and many more, it presents several difficulties that must be resolved. Several IoT features have flaws (Reshi and Sholla 2022). The IoT introduces additional security concerns since it combines networks of sensors with existing network infrastructures. Owing to its inadequate safe architecture, certain academics describe it as the Internet of Threats (Meneghello et al. 2019). As IoT adoption expands, network safety and confidentiality become more important components of an IoT framework, particularly when power, storage spaces, and the abilities of directed nodes are finite. Other reasons contributing to these devices' susceptibility include unfit Operating System (OS) abilities, proprietary software setup, placement in unorganized public surroundings, and low-end node cognitive competence. It is required to reconsider and essentially rebuild IoT systems in light of the earlier challenges. According to some researchers, "blockchain" has become the best contender tool currently available to enable a decentralized and safe environment for the IoT (Saxena et al. 2021). The idea of blockchain seems hazy since it is difficult to see how a single technology might enable functions with distinct requirements and inconsistent reliability and safety. The blockchain system now appears to be a ubiquitous technology with a wide variety of uses, not only for cryptocurrencies but also for sustaining many IoT applications while tackling their many weaknesses (Ghiro et al. 2021). There are several benefits connected to IoT systems built on blockchain technology. First, it reduces one point of breakdown, encourages resilience to faults, and permits complete communication, not relying on a centrally located and managed system. Second, consumers of a distributed ledger system may confirm both the legitimacy of the sender and the accuracy of the information being sent. Third, IoT gadgets may benefit from safe software upgrades due to blockchain's capacity for counterfeit-proof record-keeping capabilities.

Additionally, blockchain ensures provenance and tracing by preserving information and records of events in an irreversible way (Saxena et al. 2021). According to Miraz (2019), the digital ledger with the IoT could complement one another by lessening their intrinsic physical restrictions. The basic technology of the IoT is wireless and sensor networks (WSN). The IoT thus has concerns about security and privacy, just as WSN. On the other hand, blockchain technology's inherent safety, permanence, confidence, and transparency primarily fuel the movement toward its use in many industries (Hussein 2019). Blockchain technology strengthens the IoT by providing an additional layer of security, and the "things" of the Internet of Things can function as contributing nodes for blockchain technology settings, resulting in the Blockchain of Things (BCoT), a novel idea produced by fusing the two technologies. In order to increase security in general, B-IoT networks will complement one another (Hussein 2019).

## 1.11   APPLICATIONS OF THE INTERNET OF THINGS THAT UTILIZE THE BLOCKCHAIN TECHNOLOGY

Blockchain technology is being integrated into the IoT systems with innovative techniques by programmers and academics across the globe. These types of applications are primarily concerned with maximizing the advantages of blockchain's unparalleled intrinsic properties, such as permanence, reliability, the capacity to execute smart agreements, cryptographic protection, distributed management, the accuracy of data, and verification. The next part investigates numerous applications for the IoT using blockchain technology.

1.   Smart devices.
     In the years to come, all houses are expected to have smart home technologies. Particularly, it is anticipated that future homes and buildings will contain many characteristics that may be activated via smart gadgets via the Internet. Instead of recording on a central server or in the cloud and selecting blockchain ledger, confidential data can be protected, and the whole IoT system may get safety and protection.
2.   Transportation.
     Distributed ledgers in the IoT, such as tracking the origins of cars and their replacement components, are now being tested in the transportation sector. For example, Volkswagen and Jaguar Land Rover, two of the biggest automakers in the world, began investigating smart contracts with the help of IOTA, an open source ledger for IoT devices (Al_Barazanchi et al. 2022). In particular, a smart pocket is put in automobiles, enabling operators of vehicles to get IOTA tokens in return for information on traffic and roadway circumstances and even sign up for carpooling schemes. Such tokens may also be used for paying for places to park or to recharge electric automobiles.
3.   Electricity industries.
     More so than at the home or individual level, figures from intelligent devices may be utilized to reduce energy consumption across the neighborhood (Al_Barazanchi et al. 2022). Both governments and companies have already shown interest in this possibility. Blockchain technology is also expected to guarantee the continual functioning of power storage facilities.
4.   Medical care.
     Conventional client-server information management remedies are susceptible to individual sources of collapse, central information leadership, and privacy concerns. Due to its chaining mechanism and safety characteristics, blockchain technology addresses some of these underlying problems with medical systems (Saxena et al. 2021).
5.   Supply chain systems.
     Several parties working in several time regions are typically part of a supply chain, adding to its sophistication. In the past few years, several industries, including food sellers, drug companies, and electronics manufacturers, have acknowledged the value of integrating blockchain in these supply chains.

Blockchain technology is an ideal framework for ensuring an item's honesty and reliability across its entire supply chain lifecycle. It can be beneficial to have an official database that monitors the location of the source, including any changes made to these items throughout the supply chain (Saxena et al. 2021).

6. Agriculture.

There's a lot of demand for farming to expand food production healthily because of the fast rate of global population expansion. The provision of food and its use concurrently enhance people's well-being and safeguard scarce finite resources. In order to assist sustainable development for people and the planet, politicians, multilateral organizations, nonprofit organizations, and individual companies have expressed curiosity in looking into the function of the food and agricultural marketplaces. Agriculture is undergoing significant transformation and is plagued by several societal and environmental problems. The conventional agricultural techniques, which have resulted in significant forest loss, water shortages, or soil damage, cannot sustainably provide food and agriculture (Awan et al. 2021). The agricultural sector offers several potentials to expand as an effect of the IoT expansion over the past few years. The increased use of wireless broadband gadgets, smart systems, big data analysis, and artificial intelligence has given everyone involved some special capabilities for creating clear agricultural systems. Blockchain technology is one of the latest and most exciting solutions that can offer nonconventional solutions in smart farming. Blockchain technology may be utilized to more accurately regulate soils, supply chains, and storage spaces. It could be crucial for transmitting current information about animals and crops. Additionally, it may be applied to handling financial transactions, transportation, surveillance, and safeguarding food (Torky and Hassanein 2020).

7. Applications of blockchain technology in agricultural precision farming:
Utilizing the blockchain technology in precision farming may add fresh insights and enhance several processes, including effectiveness, openness, tracking, and accountability, both at the consumer and producer tiers.

## 1.12 FOOD SUPPLY CHAIN MANAGEMENT

A technique called food supply chain administration explains the way foodstuffs originate in a farm's field and then come up on our dining tables. Manufacturing, improvement, shipping, trading, utilization, and dumping are all covered by the control of the supply chain (Awan et al. 2021). Regarding the delivery of nutritious food worldwide, the secure custody of agri-foods is crucial. There could be security issues throughout the manufacturing and handling of agri-foods. As an example, excess consumption of pesticides, chemical fertilizers, or toxic metal remains from wastewater irrigation can affect the nutritional value and the security of agri-foods before and after harvesting. The contamination of subpar goods, the purposeful incorrect labeling of a food item's origin, the erroneous indication of the manufacturing and/or time of expiration, and other similar practices can all impair the purity and hygiene of agri-foods throughout manufacture. Such dangers, which represent a serious risk to the well-being of individuals, are frequently brought about by a dearth of efficient surveillance or a system of tracking (Xu et al. 2020).

Nevertheless, by establishing a distributed system and getting rid of the third-party participant, the IoT and the use of distributed ledger would solve the issue. Clients are curious about the types of foodstuffs they buy and where they come from in the traditional supply chain. The main goal herein is to pique client concerns, which is challenging given that the system is closed. The IoT plus decentralized ledger technologies may solve such issues by fostering patron trust. In this structure, each good has a digital identity that keeps information from the place of production to the merchant. Blockchain with the IoT can create an infrastructure that is further interconnected and open. Due to its distinct features, blockchain technology may provide a natural progression forward within the food supply chain that will also bring integrity by facilitating the sharing of exact information among those who are part of the supply network. The complete transparency of food goods throughout a supply chain network will become an actuality by adopting the amalgamation of the IoT and blockchain technology. Continuous surveillance and detection of authentic food products from the source and identifying key obstacles are the main benefits of the IoT and blockchain in the food supply chain.

In the same way that farmers employ harmful pesticides to spray fruits and vegetables to boost their earnings, buyers purchase the same foods to eat, posing a serious health risk. Farmers use fertilizers that contain chemicals, insect repellents, and various other substances to improve the yield of crops. Supply chain participants face significant hurdles from food theft and record manipulation (Awan et al. 2021). Considering the manufacturers' viewpoint, the usage of blockchain technology assists in building a rapport of trust with clients. It enhances the image of their goods by publicly supplying particular product data on the distributed ledger (Xiong et al. 2020).

Businesses may more effectively realize the worth of what they sell, improving how they compete. The ability of providers of dishonest and subpar goods to remain in business would be challenging, and every provider would be compelled to raise the standard of their goods across the board in the farming and food sectors. Considering the client's viewpoint, the distributed ledger facilitates access to reliable and accurate data on the production and distribution of food. It aids in addressing customers' worries about the security, excellence, and ecological sustainability of agricultural products. It comes up with a secure and perpetual means to store information gathered at the beginning of the supply chain, such as cattle species' genetics or residues of pesticides on cereals or veggies. Any entity participating in the product's supply chain can review and verify this data (Xiong et al. 2020). Figure 1.5 explains the proposed IoT-based blockchain framework for a food supply chain. Under this architecture, the primary use of the distributed ledger is for recording fresh supply chain transactions. Several IoT nodes monitor the agricultural setting and plant development. IoT devices gather information and transfer it to the interplanetary file system, which is decentralized. Thus, it contributes to the whole model's safety, scalability, and efficiency, as supply chain members also upload their data in IPFS. In a nutshell, B-IoT can greatly benefit the agriculture industry, as both promising technologies have great

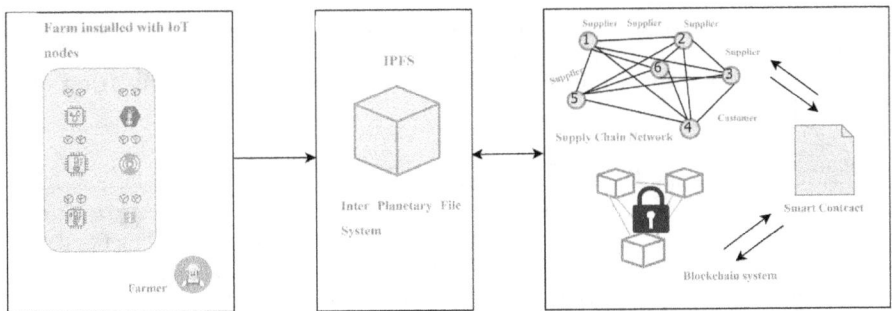

**FIGURE 1.5**　IoT-based blockchain framework for food supply chain.

potential, which complements the flaws in both and provides the agriculture industry the boon it has been yearning for a long time.

Blockchain-based technology has the power to revolutionize a wide range of businesses and sectors, but it remains in its infancy and is continually developing. Coupling IoT with distributed technology offers many opportunities and answers to most agricultural issues. Yet, numerous obstacles remain to overcome, including confidentiality and safety concerns. With other technologies like Radio Frequency Identification, B-IoT systems confront several difficulties and require additional investigation. B-IoT frameworks are subject to manipulation and assaults like Denial of Service (Dos) and Distributed Denial of Service (DDos). Also, because of their openness, such innovations lack concealment. The adoption of digital ledger is further complicated by several legal and technological obstacles, including adaptability problems that influence system performance and mining. The distributed ledger's foundation is smart contracts. However, a smart contract flaw might make the system susceptible to exterminating threats.

The B-IoT-based food supply chain has several issues that must be resolved for the B-IoT system to be broadly accepted. The majority of supply chains have a variety of partners. Thus, there needs to be a body that could keep track of them while still maintaining the decentralization of the distributed ledger. As customers are more likely to be worried about the adulteration within one item compared to the source of another, different kinds of items, such as crops, red meat, seafood, etc., might have a variety of criteria and considerations that must be considered. Consequently, for effective food SCM, each system requires various quantities of labor, time, and evaluation levels.

Even though the distributed ledger offers a trustworthy setting, it is still possible for information uploaded to the chain to be tainted and fool the entire system. Detector technology is still developing upon which the entire B-IoT system is dependent. There may be a chance that an evildoer will tamper with the sensor's technological infrastructure to gain unjust benefits. Even a fraudulent vendor could fool the network by obnoxiously positioning sensor components so that they betray the customers while

**TABLE 1.3**
**Outline of the Research That Has Been Done in the Farming Application of Blockchain Cases**

| Application | Research | Contribution |
| --- | --- | --- |
| I. Selling and buying crops. | (Umamaheswari et al. 2019) | Developed a blockchain-based IoT system to make acquiring and trading farmland and agricultural produce easier. |
| I. Food supply chain. | (Awan et al. 2021). | This paper suggested a contemporary Internet of Things integrated blockchain-based supply chain for food framework to address issues with information safety, tampering, and solitary breakdown points. |
| I. Food Supply Chain.<br>II. Agricultural Item online commerce.<br>III. Intelligent Farming. | (Xiong et al. 2020) | Analyzed the use of the technology known as blockchain from the perspective of theory and application in areas such as agricultural insurance coverage, smart agriculture, and agricultural-related product sales. |
| I. Land Registration<br>II. Supply Chain.<br>III. Food Safety.<br>IV. Farm Overseeing. | (Torky and Hassanein 2020) | Put out brand-new blockchain technology frameworks that might be utilized to address significant problems with the Internet of Things-based farming systems of all kinds. |
| I. Food Safety. | (Lin et al. 2018) | The IoT and powered by blockchain food monitoring method |
| I. Supply Chain. | (Mao et al. 2018) | System for supply chains and distribution. |

giving the supplier the bad perks without the customers even realizing that every-thing would appear to be in order as they examine the register. Additionally, these B-IoT systems are too expensive to implement. Therefore, each of the abovementioned issues might prove to be a barrier to the agriculture industry's most eagerly awaited integration of distributed ledger technology with IoT.

Table 1.3 highlights the research that was done on the farming application of blockchain cases.

## REFERENCES

101 Blockchains. n.d. "Introduction to Blockchain Features." *101 Blockchains.* https://101blockchains.com/introduction-to-blockchain-features/

Al_Barazanchi, I., A. Murthy, A. A. Al Rababah, G. Khader, H. R. Abdulshaheed, H. T. Rauf, E. Daghighi, and Y. Niu. 2022. "Blockchain Technology-Based Solutions for IOT Security." *Iraqi Journal for Computer Science and Mathematics* 3(1), pp. 53–63.

Awan, S., S. Ahmed, F. Ullah, A. Nawaz, A. Khan, M. I. Uddin, A. Alharbi, W. Alosaimi, and H. Alyami. 2021. "IoT with Blockchain: A Futuristic Approach in Agriculture and Food Supply Chain." *Wireless Communications and Mobile Computing* 2021, pp. 1–14.

Chen, Wei, Zhibin Xu, Shuang Shi, Yong Zhao, and Jianfeng Zhao. 2018. "A survey of blockchain applications in different domains." In *Proceedings of the 2018 International Conference on Blockchain Technology and Application*, pp. 17–21.

Ethereum. 2013. "A Next-Generation Smart Contract and Decentralized Application Platform." *GitHub*. https://github.com/ethereum/wiki/wiki/White-Paper

Forbes. 2018. "Will Blockchain Transform Healthcare?" *Forbes*. 5 August, 2018. Accessed May 13, 2023. www.forbes.com/sites/ciocentral/2018/08/05/will-blockchain-transform-

GeeksforGeeks. "Features of Blockchain." *GeeksforGeeks,* 25 September, 2020. www.geeksf orgeeks.org/features-of-blockchain/

GeeksforGeeks. "Introduction to Internet of Things (IoT)—Set 1." *GeeksforGeeks.* www. geeksforgeeks.org/introduction-to-internet-of-things-iot-set-1/

Ghiro, L., F. Restuccia, S. D'Oro, S. Basagni, T. Melodia, L. Maccari, and R. L. Cigno. 2021. *What Is a Blockchain? A Definition to Clarify the Role of the Blockchain in the Internet of Things.* arXiv preprint arXiv:2102.03750.

Gorbatyuk, A., and T. Gils. 2022. "Patent Transactions and the Use of Blockchain Technology." *Patent Transactions and the Use of Blockchain Technology* (September 12, 2022).

Griswold, A. "Blockchain, Explained: It Builds Trust When You Need It Most." *The Verge*, 5 October, 2021. www.theverge.com/22654785/blockchain-explained-cryptocurre ncy-what-is-stake-nft

Hussein, A. H. 2019. "Internet of Things (IOT): Research Challenges and Future Applications." *International Journal of Advanced Computer Science and Applications* 10, no. 6.

Knowledgenile. "Top 10 Applications of Blockchain Technology." *Knowledgenile*, 24 August, 2020. www.knowledgenile.com/blogs/top-10-applications-of-blockchain-technology/

Lin, J., Z. Shen, A. Zhang, and Y-C. Chai. "Blockchain and IoT Based Food Traceability for Smart Agriculture." In *Proceedings of the 3rd International Conference on Crowd Science and Engineering*, pp. 1–6. July 2018.

Mao, D., F. Wang, Z. Hao, and H. Li, 2018. Credit Evaluation System Based on Blockchain for Multiple Stakeholders in the Food Supply Chain. *International Journal of Environmental Research and Public Health*, 15(8), pp. 1627.

Meneghello, F., M. Calore, D. Zucchetto, M. Polese, and A. Zanella. 2019. "IoT: Internet of Threats? A Survey of Practical Security Vulnerabilities in Real IoT Devices." *IEEE Internet of Things Journal* 6, no. 5, pp. 8182–8201.

Miraz, M. H. 2019. "Blockchain of Things (BCoT): The Fusion of Blockchain and IoT Technologies." In *Advanced Applications of Blockchain Technology, Studies in Big Data* 60. https://doi.org/10.1007/978-981-13-8775-3_7

Nakamoto, S. 2008. "Bitcoin: A Peer-to-Peer Electronic Cash System." *Decentralized Business Review*, no. 21260.

Oracle. n.d. "What is IoT?." *Oracle*, Accessed on 13 May, 2023. www.oracle.com/in/internet-of-things/what-is-   iot/#:~:text=What%20is%20IoT%3F,and%20systems%20over%20 the%20internet

Reshi, I. A. and S. Sholla, 2022. "Challenges for Security in IoT, Emerging Solutions, and Research Directions." *International Journal of Computing and Digital Systems*, 12(1), pp.1231–1241.

Sadowsky, J. "An Introduction to Blockchain." *An Introduction to Blockchain—The CPA Journal*, 18 August, 2021. www.cpajournal.com/2021/08/18/an-introduction-to-blo ckchain/

Saxena, S., B. Bhushan, and M. A. Ahad. 2021. "Blockchain-Based Solutions to Secure IoT: Background, Integration Trends and a Way Forward. " *Journal of Network and Computer Applications* 181, pp. 103050.

Shang, B., S. Liu, S. Lu, Y. Yi, W. Shi, and L. Liu. 2020. "A Cross-Layer Optimization Framework for Distributed Computing in IoT Networks." *IEEE/ACM Symposium on Edge Computing (SEC)*, pp. 440–444.

Singh, B. P., and A. K. Tripathi. 2019. "Blockchain Technology and Intellectual Property Rights." *Journal of Intellectual Property Rights* 24(4), pp. 269–276.

Singhal, B., G. Dhameja, and P. S. Panda. 2018. "How Blockchain Works." In *Beginning Blockchain: A Beginner's Guide to Building Blockchain Solutions*, pp. 31–148. New York: Apress.

Torky, M., and A. E. Hassanein. 2020. "Integrating Blockchain and the Internet of Things in Precision Agriculture: Analysis, Opportunities, and Challenges." *Computers and Electronics in Agriculture* 178, pp. 105476.

Umamaheswari, S., S. Sreeram, N. Kritika, and D.J. Prasanth. 2019. "Biot: Blockchain Based IoT for Agriculture." In *2019 11th International Conference on Advanced Computing (ICoAC)*, pp. 324–327. IEEE. Chennai, India.

Wang, J., S. Wang, J. Guo, Y. Du, S. Cheng, and X. Li. 2019. "A Summary of Research on Blockchain in the Field of Intellectual Property." *Procedia Computer Science* 147, pp. 191–197.

Wüst, K., and A. Gervais. 2018, June. "Do You Need a Blockchain?" In *2018 Crypto Valley Conference on Blockchain Technology (CVCBT)*, pp. 45–54. IEEE. Zug, Switzerland.

Xiong, H., T. Dalhaus, P. Wang, and J. Huang. 2020. "Blockchain Technology for Agriculture: Applications and Rationale." *Frontiers in Blockchain* 3, pp. 7.

Xu, J., S. Guo, D. Xie, and Y. Yan. 2020. "Blockchain: A New Safeguard for Agri-foods." *Artificial Intelligence in Agriculture* 4, pp. 153–161.

# 2 Blockchains in IoT
## Introduction, Features, and Vulnerabilities

*Neeraj Gupta*

Panipat Institute of Engineering and Technology, Haryana, India

## 2.1 INTRODUCTION: BACKGROUND AND DRIVING FORCES

Kevin Ashton coined the word "Internet of Things" (IoT) to describe a system where physical objects can collaborate using the Internet. The principal activities involved in IoT are data acquisition, data collection, data transportation and data storage, data processing, and actuation. To support such a large array of activities and manage a vast repository of data requires the support of sensors, communication protocols, cloud computing, data analytics, and actuators to meet that desired outcomes. The five data characteristics in IoT systems are volume, velocity, variety, variability, and veracity. The application of IoT in industries for automation is popularly termed the Industrial Internet of Things (IIoT). The IoT applications range from manufacturing, healthcare, transportation, education, finance, and many others. Statista forecasted that by 2030, around 29 billion devices will be deployed worldwide. Privacy, counterfeit hardware, secure data access, secure data sharing, scalability, and reliability are some security challenges that IoT systems face.

Blockchains are decentralized, distributed, shared, and immutable ledgers that facilitate transaction recording, validation of transactions, and tracking of resources. The resources can be tangible and intangible assets. The cryptographic hash, digital signatures, and consensus algorithms make blockchains secure, decentralized, and distributed. Peers can be incentivized to act as good actors in networks to ensure the stability of a decentralized system. Cryptoeconomics is a multidisciplinary area of computer science that aims to solve the coordination problem between various participants through cryptography and economics. Rather than just a subset of economics, it involves a mix of game theory, mathematics, distributed computing, and software engineering. Blockchains gained popularity due to cryptocurrencies, but in the last decade, many corporates and governments have realized the substantial business prospects of this technology. The 2030 blockchain market share is forecasted to be 1,235.71 billion US dollars. Various features of blockchains are suitably adapted to the security-related problem associated with IoT. The amalgamation of two popular technologies, blockchain and IoT, is called BIoT, which has new advantages and challenges. The current chapter identifies the architecture of BIoT, security issues solved through the adaption of blockchains in IoT, various consensus algorithms, and other cases reported in the literature. The second section discusses the IoT architecture and challenges faced by the IoT systems. The third section describes the working and key terms associated with blockchain. The fourth section discusses the key features of blockchains in IoT. Section five discusses various platforms available in the marketplace that have adopted blockchain for developing IoT applications. Section six illustrates different challenges in B-IoT. These challenges pave the way for crucial research. Section seven gives direction into key research areas that are open for further exploration. Section eight summarizes the chapter.

## 2.2   INTERNET OF THINGS

IoT can be defined as the interconnection of physical devices or sensors that are referred to as physical objects that communicate through the Internet, primarily wireless communication protocols, to interact with their internal or external states. Gartner defines IoT platforms as follows: "An IoT platform is an on-premises software suite or a cloud service (IoT platform as a service [PaaS]) that monitors and may manage and control various types of endpoints, often via applications business units deploy on the platform" (IoT Platforms Reviews 2023 I Gartner Peer Insights).

The four essential characteristics of the IoT are the following:

1.   They should be efficient and scalable.
2.   They should have unambiguous naming and address convention.
3.   There can be large sleeping nodes, mobiles, and non-IP devices.
4.   There could be intermittent connectivity between source and destination.

IoT architecture has layered architecture, just like traditional computer networks. Primarily there are two prominent architectures: (a) one Machine to Machine (M2M) architecture ("OneM2M Overview – OneM2M" n.d.) and IoT world forum reference architecture (Atlam, Walters, and Wills 2018). The scale and heterogeneity of devices in IoT networks led to specific challenges that need to be addressed. The authors (Cui et al. 2019) have appropriately divided these challenges into five categories. Figure 2.1 illustrates various taxonomies concerning IoT challenges.

1.   System Design: The present IoT framework is based on a centralized architecture. The data collected from the objects is collected at a central cloud storage device. This data is subjected to data analytics and machine learning algorithms to ensure the decision-making process can be automated. The process must fulfill the critical service level criteria, including throughput, latency, quality of service, redundancy, and security. Deployment of many IoT devices necessities calls for decentralized solutions.
2.   Data Management: The five V's in IoT need special attention since they introduce various challenges. The requirement to store voluminous data and algorithms required to clean and process this data add to the carbon prints of the data centers. The security issues attached to data are another important factor that needs attention.

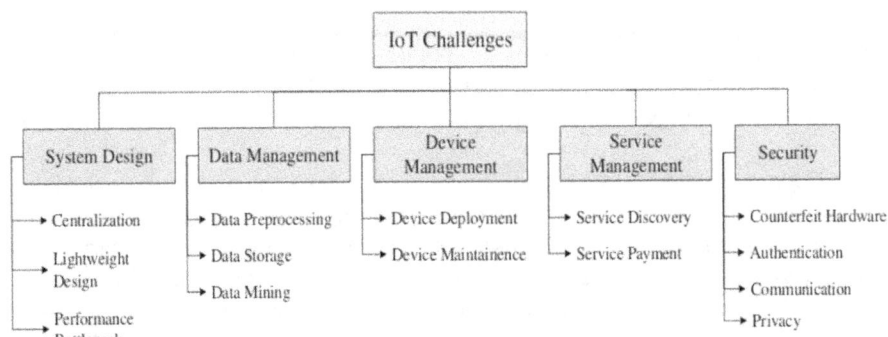

**FIGURE 2.1** Taxonomy of IoT challenges.

**Source: Cui et al. 2019.**

3. Device Management: To conserve devices' energy, they must operate in localized wireless environments. Although many lightweight data and communication protocols are proposed, the issue of security, reliability, and efficiency pose various challenges that need to be addressed. The heterogeneity in communication protocols and addressing schemes is a further challenge.

4. Service Management: Cloud computing promotes the idea of everything as a service. This concept broadly includes hardware resources, networks, middleware, and applications. Microservices architecture facilitates the interaction with the resource and, at the same avoids a single point of failure. It is essential to ensure that devices can automatically discover services transparently and reliably. The devices should be able to compensate each other for using the service without human intervention.

5. Security: Advances in technology, including artificial intelligence (AI), have provided hackers with a new set of tools that are more lethal and can lead to substantial monetary and infrastructure losses to individuals, business groups, and governments. The IoT device is low-powered, has limited processing capacity, and has low memory capacity, which exposes them to various online threats. The security problem in IoT can be divided into authentication, connection, and operation (ŞİMŞEK 2021). More sophisticated techniques like counterfeit hardware and microarchitectural attacks (Naghibijouybari, Koruyeh, and Abu-Ghazaleh 2022) are used to target IoT infrastructures.

## 2.3 BLOCKCHAINS

Blockchains are distributed ledgers that work on peer-to-peer networks enabling the recording of transactions using a consensus algorithm. The blockchains are based on distributed ledger technology (DLT) which permits participants in a decentralized network to record, exchange, synchronize, and validate transactions ("DLT: The Future Is Distributed – KPMG Global," 2022). DLT uses cryptographic algorithms to secure assets transparently and trustworthily. There are two kinds of DLTs: blockchains and directed acyclic graphs. Blockchain was introduced by Satoshi Nakamoto in 2008 through his work on Bitcoin. The popularity of Bitcoins paved the way for more cryptocurrency. The following properties of blockchain encourage adoption in various other sectors (Abdelmaboud et al. 2022):

1. Decentralization: Unlike traditional ledgers, blockchains distribute the ledger across the entire network without requiring a mediator. The technology maintains multiple replicas like peer-to-peer (P2P) torrent file sharing. Any transaction constituting the block is added to the chain of the blocks only if most of the peers reach a common point of agreement about the state of the ledger contents and updates. The deal is achieved through consensus algorithms such as proof of work and proof of stake (Cui et al. 2019). The consensus algorithms should ensure fairness and security.

2. Immutability: Blockchains guarantee that every block, once added, is immutable. The immutability is achieved through a hash function. The block is rehashed and timestamped to maintain sanity and immutability. Every block has three basic elements: data, nonce, and hash value. A nonce value is a 32-bit pseudo-random number generated during the mining process and primarily used as a counter in the chain. The nonce serves as a promising tool in authentication protocols for safeguarding old communications from reprocessing. Some e-signature tools also use blockchain nonce for creating, comparing, and verifying signatures. To further maintain the immutability property in the decentralized network, each block contains its hash address and the hash address of its previous block.

3. Transparency: Blockchain builds transparency based on the asymmetric algorithm. Each node in the network is assigned two keys: public key and private key. The public key identifies the other user with whom the transaction is to be accomplished. Anybody in the network can use the public key and derive the address through one-way cryptographic functions. The public keys can also access the transaction made in the network. The private keys, on the other hand, are used to confirm ownership and control digital assets. Private keys create digital signatures of users. Each transaction made by the user is digitally signed to ensure the integrity and authentication of the transactions (Thakore et al. 2019). Miners based on a consensus algorithm validate the transaction between two parties. Once the transaction is validated, the block (the whole chain of blocks) is open and available for open audit. Interestingly, the participant's public key is derived from its private key using the function elliptic curve multiplication. These algorithms are one-way in nature, i.e., it is easy to generate the public keys, but it is impossible to regenerate the private key from it. Figure 2.2 demonstrates the working methodology of blockchains.

4. Trust: Merkel trees is a data structure, essentially a binary hash tree, that simplifies the summarization and validation of large data sets. Each transaction in the blockchain generates a hash value for a block. In the Merkel tree, the hash code of the

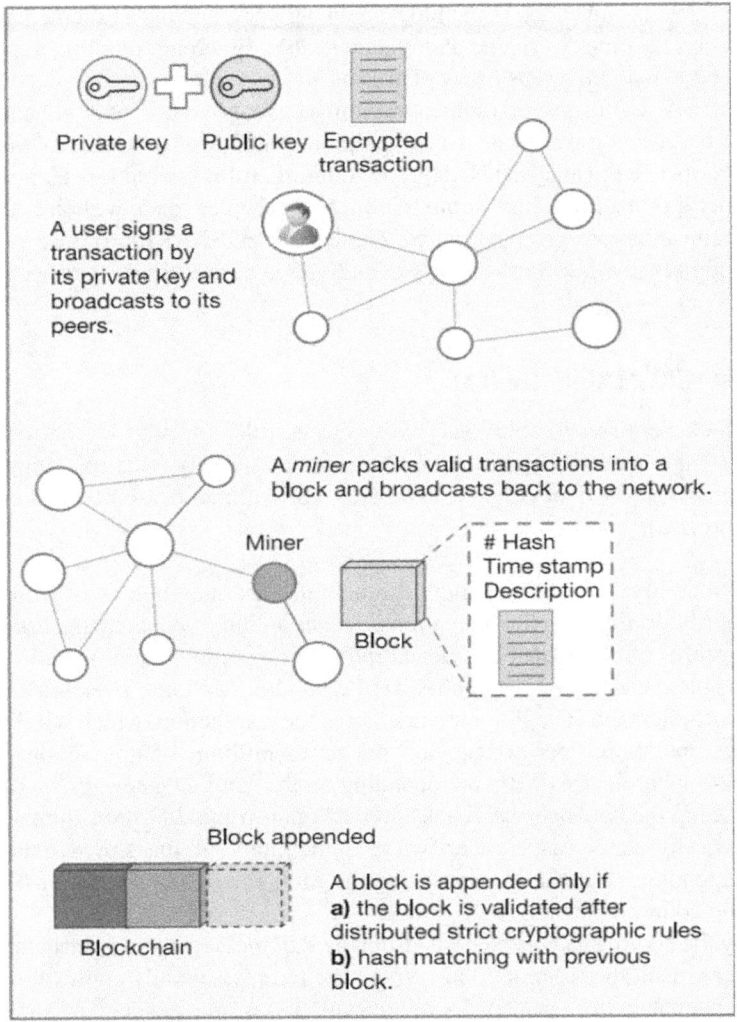

Private key    Public key  Encrypted
                           transaction

A user signs a
transaction by
its private key and
broadcasts to its
peers.

A *miner* packs valid transactions into a
block and broadcasts back to the network.

Miner

# Hash
Time stamp
Description

Block

Block appended

Blockchain

A block is appended only if
**a)** the block is validated after
distributed strict cryptographic rules
**b)** hash matching with previous
block.

**FIGURE 2.2**   Blockchain working methodology.

**Source: Ferrag et al. 2019.**

block labels each leaf node. All non-leaf nodes contain the cryptographic
hash of all labels of their children. The process is iteratively followed until
only one node is left; this node is called Merkel Node or root node. The user
interested in verification can do the same by providing only the hash value of
the transaction. This process helps save computing resources in processing
and storage for verification purposes. The verification time is proportional
to the number of leaf nodes in the tree. Another way of implementing trust
in blockchains is through smart contracts. Smart contracts are automated
computer scripts executed when parties agree on specific requirements. The

contract is embedded in the blockchain network, and once executed, the transaction becomes trackable and unchangeable. Ethereum platform supports the use of smart contracts on its platform.

5. Privacy: Using cryptographic algorithms ensures the immutability and privacy of the user's private data. Cryptographic techniques such as Zero Knowledge Proofs (ZKPs) and zk-SNARKs use homomorphic encryption. Homomorphic encryption allows the computation of encrypted data without decrypting them. A crypto-protocol called Zcash uses zk-SNARKs to encrypt its data and only gives decryption keys to authorized parties for them to see that data.

## 2.4  BLOCKCHAINS IN IOT

Many blockchain-based solutions have successfully managed, controlled, and secured IoT applications. Apart from addressing the security issues plaguing the IoT, blockchains ensure IoT devices' scalability, security, interoperability, and orchestration (Figure 2.3).

1. Scalability: Low latency, high throughput, fast querying, and permission are highly desired in IoT applications. The adoption of blockchains brings in the features of distributed and decentralized computing which will eliminate the problem associated with centralized computing and improves scalability and fault tolerance. The P2P network stores the transaction, which will help scale the operation. Peer-to-peer network eases millions of transactions done by billions of devices without depending on the central "gateways," – thus eliminating the network overheads and the "man-in-middle" procedure.

2. Security: It is easy to verify the authenticity of the stored data without depending on the third party using the Merkel tree and timestamps. The data and addresses shared
   by the nodes are encrypted, securing the P2P messaging and transmission. The consensus algorithm used in cryptocurrencies, essentially public blockchains, can enable the user to do microtransactions independent of its location. Private blockchains will allow the data to be handled securely and transparently. The administrator in the private chain can restrict the number of users and control the privileges given to users. The administrator can easily track any attempt made to change the ledger. Solutions based on secure multiparty computing, zero information checks, commitment plans, ring signatures, and homomorphic hiding ensure privacy security (Bernal et al. 2019).

3. Interoperability: Interoperability should allow data exchange between two or more IoT systems efficiently and effectively. It is of paramount importance that data integrity and authenticity are always maintained. While defining the architecture of the IoT system, the developer can choose from various blockchain platforms depending on the need and requirement of the application. Each blockchain platform utilizes different consensus algorithms. The B-IoT entities or applications belonging to the same blockchain network

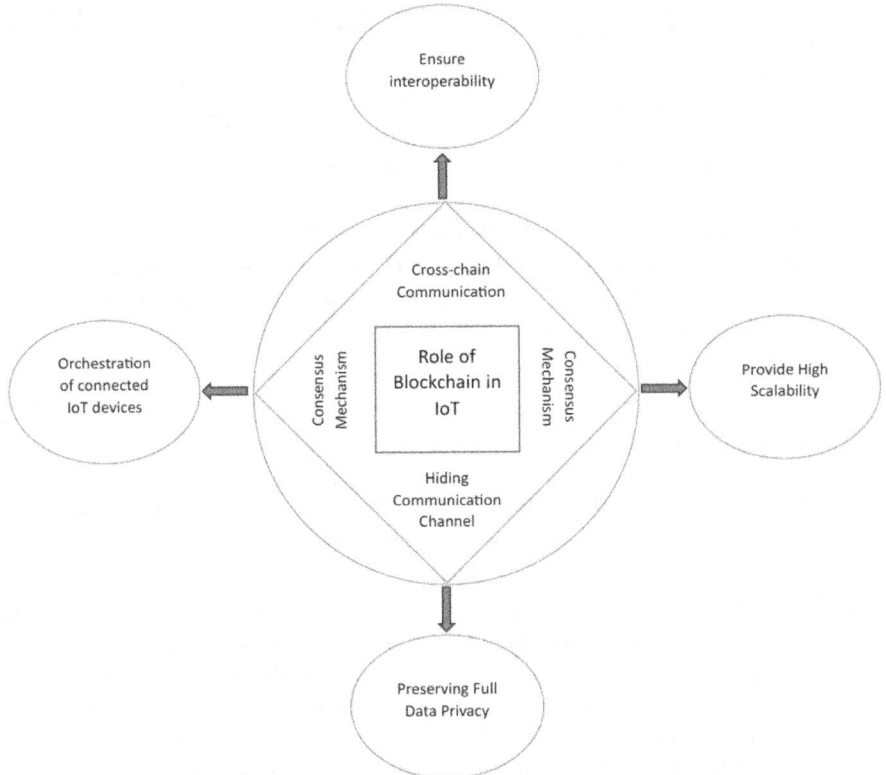

**FIGURE 2.3**   Role of blockchain in IoT.

**Source: Abdelmaboud et al. 2022.**

can communicate. The lack of inter-blockchain or cross-communication will hinder technology adoption in other fields. A blockchain bridge facilitates the transfer of assets between two chains. Different rules, protocols, and structures may govern these chains. The bridges thus assist in cross-communication between two different blockchains. The main advantages of bridges are low network traffic, better decentralized apps, and inhibition of monopolization by a particular blockchain platform.

There are various ways and approaches to simplify the transaction between two heterogeneous blockchains without involving any third party: atomic swaps, relays, merged consensus, federation, and stateless simplified payment verification. In nuclear swaps, the micro-transaction will include the exchange of tokens in a peer-to-peer network. Thus, it must be considered as something other than an accurate cross-communication method. Relays involve using smart contracts to validate the other chain's transactions, events, and block headers. Merged consensus provides two-way interoperability between chains through chains. Federations allow trustworthy groups to validate occurrences on one chain on another.

4.  Orchestration of IoT devices: The management of IoT infrastructure can be managed effectually using smart contracts. They can be written to make decisions concerning management and access of data, data development and processing, dynamic positioning, and security concerns. The IoT allocates unique identification to each device in the network. Similarly, each block entered into the system can be uniquely identified by the hash code, which can be easily verified. Thus, blockchain empowers the IoT system to work autonomously. Adopting blockchains in IoT will also help reduce cost, accelerate data exchange, scale security, build trust, and eliminate single points of failure (Gopal 2016).

## 2.5  B-IOT PLATFORMS

Integration of the IoT and blockchains has led to the development of various B-IoT platforms.

1.  IoTChain: Alphand et al. (2018) proposed IoTChain as a combination of the authorization in constrained environments (ACE) framework and the Object Security Architecture for the Internet of Things (OSCAR) architecture. It is a three-tier architectural security platform comprising an authentication layer, a blockchain layer, and an application layer. The main pros of the proposed architecture are identity authentication, access control, privacy protection, lightweight feature, regional node fault tolerance, denial-of-service resilience, and storage integrity. "Metanode") runs on IOTChain and enables the users to deploy D-Apps on their mobiles.
2.  IOTA: IOTA is a third-generation DLT technology well suited for the IoT. The main features of IOTA are block less, feeless, permissionless, mining-less, secured, and scalable protocol. It introduces the new architecture, "Tangle," which enables feeless micro-transaction between multiple devices. Essential IOTA is based on DAG rather than blockchains. It is a public permissionless distributed ledger that is low on resource requirements and fast on transactions. Each participant who wants to do a transaction must validate two previously confirmed transactions. It has no miners, and all nodes are equal participants and validators. They use quantum-resistant cryptographic algorithms which are immune to various attacks.
3.  IExec: They are the first to offer a blockchain-based marketplace ("IExec Technical Documentation – IExec Doc" n.d.). The platform utilizes Ethereum smart contracts to provide high-performance cloud infrastructure on demand. The publisher-subscriber model facilitates the exchange of services, data sets, and computing services. It employs proof-of-contribution consensus protocols, which also support offline validations.
4.  Xage: The objective is to secure the IIoT data by deploying blockchain in its security product. The platform uses a "supermajority" consensus algorithm

and a hierarchical tree to support blockchain synchronization. One of the key features is the use of identity-based access control to protect users, machines, apps, and data, at the edge and in the center/cloud. The data is validated first at the local level and then at the global level. These operations ensure that local transactions can continue even if the regional nodes disconnect from the global network.

5. SONM: It is a blockchain platform based on Ethereum technology but charges no fees for gas. The platform simplifies the development of decentralized fog-based applications. Ethereum-based smart contracts facilitate the financial-based apps that are deployed that are responsible for order processing, deal processing, and payments.

6. UniquID: The platform is open-source authentication and authorization identity access management for IoT devices working under unreliable connectivity (Giaretta, Pepe, and Dragoni 2019). It works on the application layer; the protocol is modular and atomic and uses micro-transactions to authenticate and authorize network endpoints. The protocol can be efficiently deployed on gateways, routers, virtualized servers, containers, mobile phones, and serverless applications. Post November 2021, the protocol is available for non-commercial use and experimentation.

7. TraceRx: The drug supply chain management suffers from poor recalling, stock management, and theft of medicines. TraceRx ("Tracerx: Global Blockchain Supply Chain for Drugs" n.d.) is a platform that solves the above problems. United Nations Organizations (UNO) is currently using the platform to trace the distribution of free drugs and identify inefficiency and losses. The admin can connect any of the nodes and trace the shipment status.

8. Helium: It is a blockchain-based IoT company that enables connectivity between devices across the globe without depending on satellite connections and expensive cellular infrastructure. The main idea is to convert the centralized physical networks into a wide-area decentralized networking system. The network is wireless and based on blockchains and the helium wireless protocol. The blockchain runs on the new consensus protocol: proof-of-coverage. The miners must authenticate their location and time to be added to the Helium network. The main characteristics of the networks are the following: (a) devices need to pay for sending and receiving their data and geolocation; (b) the miners added into the Helium network earn tokens to provide network coverage; and (c) miners can charge transaction fees for validating the integrity of the network (Haleem et al. 2018).

9. NetObjex: The platform provides a standard and decentralized network for communication among various IoT devices. The blockchain-based IoTokens facilitate communication among IoT devices belonging to the same ecosystem. The tokens can be fungible to non-fungible assets stored in digital wallets. The stored tokens in wallets can interact with decentralized applications on various blockchains ("Home – The Platform for NFT MarketPlace and Web3 Wallet" n.d.).

10. RDDL: Web3 applications are based on blockchains and are decentralized. The RDDL network works on unique Proof of Productivity consensus algorithm. The network provides machines with unique, incorruptible identities, equipping them with all the necessary tools to communicate and transact securely and autonomously via smart contracts and within incentive networks (Fürstner 2022). The network aims to provide a common infrastructure allowing machines to interact directly and autonomously with physical assets with Web3 technology. Some of the applications envisioned by its creator are (a) the creation of a machine data economy that can accelerate the emergence of smart cities, smart supply chains, smart factories, etc.; (b) the adoption of renewable energy through financing using token incentives that were generated while deploying infrastructure; and (c) introduction of machine-centric NFT propositions and services based around digital twins.

## 2.6 B-IOT: CHALLENGES

This section discusses various challenges that need attention. Solutions to these challenges will help in realizing the true potential of B-IoT.

1. Resource Constraints: Blockchains' mining and verification processes are power-hungry and computation-intensive, and require huge storage requirements. Most IoT devices have limited power, simple hardware specification, and restricted processing power. This limitation needs to improve transaction times and high scalability. There has been an effort to offload the computations to the cloud, but this will increase latencies. The consensus process is time-consuming since it requires users to reach an agreement to maintain the correct global state of the chain. Due to the huge amount of data generated by IoT devices, this process would require high bandwidth (Lao et al. 2020).

2. Security: Integrity, confidentiality, and availability are three pillars of security. It is an absolute requirement to ensure that the data produced by billions of devices is secure. Vulnerabilities have been reported for the elliptic curve digital signature algorithm to generate private keys. This vulnerability will expose the chain to various attacks. Most attacks can be divided into two categories: (i) application-dependent and (ii) application-free attacks (Ferrag et al. 2019). The proposed security model suggested in the literature addresses only the subset of the security concerns for application-free attacks. It is to identify the vulnerability of a specific application. Countermeasure can be incorporated to repel any particular threat. Designing a resilient security solution against most attacks is crucial, considering that IoT devices are resource constrained. Heterogeneity in IoT networks demands an adaptable and dynamic security framework for B-IoT.

3. Data set availability and sharing: The evaluation of security protocols can better be judged on the actual data rather than the simulated data sets. However, relatively few real-world data sets are available in the assessment by

researchers (Banerjee, Lee, and Choo 2018). The availability of standardized data sets is much desired, but at the same time, it is crucial to maintain the privacy and integrity of such data sets.

4. Legal Regulations: Although blockchains are used widely in various applications, there needs to be a more legal and regulatory understanding of the issues concerning breach of privacy and data leakage. Cryptocurrencies are illegal tenders in many countries across the globe. It is essential to undertake regulatory initiatives such as name anonymization, issue of accessibility and termination rights, issues of openness, breach of privacy, and administrative and local governance.

5. Integration of IoT Communication Protocols: It is essential that IoT communication protocols like Bluetooth, IEEE802.15.4, Zigbee, 6LoWPAN, LoRaWAN, etc., are integrated with blockchains for the transaction of records, verification, and monetization.

6. Software Upgrades: It is important to continuously upgrade the firmware/software of IoT devices to insulate them from cyber attacks. The decentralized architecture of blockchains makes it difficult and complex to execute the synchronized software upgrades of end-IoT devices.

7. Scalability: Scalability is a major deterrent in adopting blockchains in IoT. The fully participating node in the blockchain is required to keep a copy of the data. As sensors increase, the data generated will grow exponentially in IIoT. The prolongation of time required to process the transaction will be a problem. Despite lightweight algorithms like Ethereum, there will be huge storage and resource constraints (Alladi et al. 2019). In another architecture proposed by Lao et al. (2020), IoT devices are neither required to store a copy of the blockchain nor conduct mining work. These nodes are thus lightweight. There are preconfigured nodes in the backbone structure where all the heavy computation work is allocated. The IoT nodes are required to cooperate with these nodes. Any malfunction of an IoT device does not adversely affect the network performance.

8. High Cost: Maintaining blockchains networks across heterogeneous networks is computationally expensive. The cost can be accounted for by computational power, energy consumption, and memory required to process the data. The expansion of IoT will generate volumes of data leading to low transaction rates. In recent years there has been a proposal to offload the computation from fog devices by utilizing cloud servers or fog servers; this, however, will increase the network latencies.

9. Smart-Contract-Related Challenges: A contract is a piece of code that runs in a blockchain network, validating the transaction if certain conditions are verified. These intelligent contacts can be integrated into IoT systems to authorize and authenticate devices and warrant application interactions and security. Smart contracts rely on the data fed up by real-world systems through oracles. Oracles are middleware that eliminates intermediaries when providing data to smart contracts. The oracles are susceptible to manipulation, corruption, and error.

## 2.7   B-IOT: RESEARCH DIRECTIONS

This section outlines various research directions and pointers that would improve the effectiveness and capability of blockchains in various IoT applications.

a.   The intelligence in IoT can be introduced at various layers starting from the device to the cloud level. Similarly, extensive work has been done in "AI-driven blockchains" and "blockchain-driven AI." It is important to merge AI-based machine learning algorithms, blockchains, and IoT systems to work in tandem to identify the threat and take necessary preventive action in an automated manner. There is much scope for working in predictive AI-based B-IoT.

b.   Lightweight Consensus Algorithms and Smart Contracts: IoT devices are primarily resource constrained. As discussed earlier, heavyweight consensus algorithms can be resource-intensive and unsuitable for many IoT applications. The development of lightweight self-healing blockchain algorithms is one area that needs further exploration. Such algorithms can be readily deployed on intelligent devices enabling the user's trust.

c.   Emerging Threat Landscapes: Security has always been a cat–mouse show, and emerging threats and vulnerabilities need constant monitoring. It will be interesting to see how blockchains monitor various hardware-based and software-based security threats. Banerjee, Lee, and Choo (2018) have pointed out an attractive research area where investigators will obtain evidential data from encrypted communications where the investigators and incident responders do not have access to the decryption key.

d.   Standardize Test Platforms: Before deployment, BIoT applications should be tested for stability, performance, and security. Following salient inferences were drawn by Imperius and Alahmar (2022) in their review concerning testing strategies for smart contracts.

   • Many new tools, validation processes, and metrics are being proposed, but these proposals need more evaluation and validation. Process validation in this active domain area
     needs attention.
   • Ethereum network is the most preferred platform for developing applications employing smart contracts. However, other platforms also incorporate smart contracts. Applications must be able to support cross-transactions. There needs to be more testing procedures for such applications.
   • There is a requirement to create robust benchmarks for testing smart contracts.
   • Although new testing methods are being proposed, more data must be used to compare various testing strategies. There needs to be more knowledge in this domain.

e.   Trust: The IoT devices generate massive data which will feed up to blockchains. Interdomain policies and control systems must be maintained to develop confidence in BIoT implementations. There is a need to pursue more research in this area.

f. Embedded Systems: Another promising area of research is designing embedded systems, including blockchain-based hardware and blockchain-oriented software ad platforms. The design of such a system will promote the seamless integration of blockchains and IoT systems.

g. Integration: Another research direction is the integration of blockchains along with modern cryptographic algorithms and fog computing for developing IoT applications.

## 2.8 CONCLUSION

The IOT is pivotal in industries, healthcare, transportation, space exploration, home application, agriculture, defense, and pollution control. Numerous IoT devices pump data for storage in edge devices and cloud environments. Security of this voluminous data is one of the primary concerns actively pursued by academia and industries alike. The adoption of blockchains for providing protection is in its nascent stage. Many vital areas need to be researched for blockchains before cyber-physical systems fully adopt them. This chapter reviewed the working of IoT and blockchains illustrating the key benefits that can be harvested by adopting blockchains in IoT. Some of the popular B-IoT platforms for the development of applications are discussed in Section 2.5. Many issues and challenges are open, requiring a lot of thought and deliberation for more secure and transparent usage of blockchains in IoT applications.

## REFERENCES

Abdelmaboud, Abdelzahir, Abdelmuttlib Ibrahim Abdalla Ahmed, Mohammed Abaker, Taiseer Abdalla Elfadil Eisa, Hashim Albasheer, Sara Abdelwahab Ghorashi, and Faten Khalid Karim. 2022. "Blockchain for IoT Applications: Taxonomy, Platforms, Recent Advances, Challenges and Future Research Directions." *Electronics (Switzerland)* 11 (4): 1–35. https://doi.org/10.3390/electronics11040630

Alladi, Tejasvi, Vinay Chamola, Reza M. Parizi, and Kim Kwang Raymond Choo. 2019. "Blockchain Applications for Industry 4.0 and Industrial IoT: A Review." *IEEE Access* 7: 176935–51. https://doi.org/10.1109/ACCESS.2019.2956748

Atlam, Hany F., Robert J. Walters, and Gary B. Wills. 2018. "Internet of Things: State-of-the-Art, Challenges, Applications, and Open Issues." *International Journal of Intelligent Computing Research* 9 (3): 928–38. https://doi.org/10.20533/ijicr.2042.4655.2018.0112

Banerjee, Mandrita, Junghee Lee, and Kim Kwang Raymond Choo. 2018. "A Blockchain Future for Internet of Things Security: A Position Paper." *Digital Communications and Networks* 4 (3): 149–60. https://doi.org/10.1016/j.dcan.2017.10.006

Bernabe, Jorge Bernal, Jose Luis Canovas, Jose L. Hernandez-Ramos, Rafael Torres Moreno, and Antonio Skarmeta. 2019. "Privacy-Preserving Solutions for Blockchain: Review and Challenges." *IEEE Access* 7: 164908–40. https://doi.org/10.1109/ACCESS.2019.2950872

Cui, Pinchen, Ujjwal Guin, Anthony Skjellum, and David Umphress. 2019. "Blockchain in IoT: Current Trends, Challenges, and Future Roadmap." *Journal of Hardware and Systems Security* 3 (4): 338–64. https://doi.org/10.1007/s41635-019-00079-5

"DLT: The Future Is Distributed – KPMG Global." 2022, March. https://kpmg.com/xx/en/ home/insights/2022/01/dlt-the-future-is-distributed.html

Ferrag, Mohamed Amine, Makhlouf Derdour, Mithun Mukherjee, Abdelouahid Derhab, Leandros Maglaras, and Helge Janicke. 2019. "Blockchain Technologies for the Internet of Things: Research Issues and Challenges." *IEEE Internet of Things Journal* 6 (2): 2188–2204. https://doi.org/10.1109/JIOT.2018.2882794

Fürstner, Tom. 2022. "The RDDL Network The Physical Trust Stack for Web3." Vol. 3.

Giaretta, Alberto, Stefano Pepe, and Nicola Dragoni. 2019. "UniquID: A Quest to Reconcile Identity Access Management and the IoT." *Lecture Notes in Computer Science (Including Subseries Lecture Notes in Artificial Intelligence and Lecture Notes in Bioinformatics)* 11771 LNCS: 237–51. https://doi.org/10.1007/978-3-030-29852-4_20

Gopal, S. 2016. "Blockchain for the Internet of Things." www.tcs.com/blockchain-for- iot

Haleem, Amir, Andrew Allen, Andrew Thompson, Marc Nijdam, and Rahul Garg. 2018. "Helium A Decentralized Wireless Network." Vol. 2. https://github.com/helium

"Home – The Platform for NFT MarketPlace and Web3 Wallet." n.d. Accessed June 25, 2023. www.netobjex.com/

"IExec Technical Documentation – IExec Doc." n.d. Accessed May 3, 2023. https://docs.iex.ec/

Imperius, Nicholas Paul, and Ayman Diyab Alahmar. 2022. "Systematic Mapping of Testing Smart Contracts for Blockchain Applications." *IEEE Access* 10 (October): 112845–57. https://doi.org/10.1109/ACCESS.2022.3216874

"IoT Platforms Reviews 2023 I Gartner Peer Insights." n.d. Accessed April 29, 2023. www.gart ner.com/reviews/market/iot-platforms

Lao, Laphou, Zecheng Li, Songlin Hou, Bin Xiao, Songtao Guo, and Yuanyuan Yang. 2020. "A Survey of IoT Applications in Blockchain Systems: Architecture, Consensus, and Traffic Modeling." *ACM Computing Surveys* 53 (1). https://doi.org/10.1145/3372136

"Metanode I IoT Chain Platform Is the World's First Layer 0 Blockchain on Mobile." n.d. Accessed May 3, 2023. https://metanode.co/

Naghibijouybari, Hoda, Esmaeil Mohammadian Koruyeh, and Nael Abu-Ghazaleh. 2022. "Microarchitectural Attacks in Heterogeneous Systems: A Survey I ACM Computing Surveys." *ACM Computing Surveys.* 2022. https://doi.org/https://doi.org/10.1145/ 3544102

"OneM2M Overview – OneM2M." n.d. Accessed May 5, 2022. https://wiki.onem2m.org/ index.php?title=OneM2M_overview

ŞİMŞEK, Mehmet Ali. 2021. "A Study of Blockchain in IoT Architecture." *International Journal of Engineering and Innovative Research*, May. https://doi.org/10.47933/ ijeir.851109

Thakore, Riya, Rajkumar Vaghashiya, Chintan Patel, and Nishant Doshi. 2019. "Blockchain – Based IoT: A Survey." *Procedia Computer Science* 155: 704–9. https://doi.org/10.1016/ j.procs.2019.08.101

"Tracerx: Global Blockchain Supply Chain for Drugs." n.d. Accessed June 24, 2023. www. leewayhertz.com/project/tracerx/

# 3 Blockchain-Based Internet of Things (B-IoT)

## Challenges, Solutions, Opportunities, Open Research Questions, and Future Trends

*Wasswa Shafik*

School of Digital Science, Universiti Brunei Darussalam, Gadong, Brunei Darussalam, and Dig Connectivity Research Laboratory (DCRLab), Kampala, Uganda.

## 3.1 INTRODUCTION

The convergence of blockchain technology and the Internet of Things (IoT) has produced a new blockchain-based Internet of Things (B-IoT) paradigm. B-IoT holds tremendous potential for revolutionizing various industries and domains by providing enhanced security, trust, and decentralized data management (Han et al., 2023). In B-IoT, the blockchain acts as an immutable and transparent distributed ledger, enabling secure and efficient transactions between IoT devices (Issa et al., 2023). Integrating blockchain with IoT introduces a layer of transparency, immutability, and consensus that can address the challenges associated with data integrity, trust, and security in IoT ecosystems (Sasikumar et al., 2023). By leveraging cryptographic techniques and decentralized consensus mechanisms, B-IoT enables verifiable and auditable interactions between connected devices, facilitating the creation of decentralized, autonomous systems (Shafik, Matinkhah, Asadi, et al. 2020; Tyagi et al. 2023).

The opportunities presented by B-IoT are vast and diverse. For example, the enhanced security and transparency offered by blockchain can greatly benefit critical sectors such as healthcare, where secure sharing and access to medical records can improve patient care and enable interoperability among healthcare providers (Huang, Yang, and Ajay, 2023; Shafik, Matinkhah, and Ghasemazade, 2019a). Similarly, in supply chain management, B-IoT can ensure traceability and authenticity, enabling efficient tracking of goods, reducing fraud, and improving the efficiency of supply chain processes (Yang et al., 2021). Furthermore, in energy systems, B-IoT can enable peer-to-peer energy trading, decentralized grid management, and transparent tracking of renewable energy generation (Shafik et al., 2021).

However, integrating blockchain and IoT also brings unique challenges that must be overcome for successful implementation and widespread adoption. Scalability remains a crucial concern, as the massive volume of IoT data generated requires efficient consensus algorithms and storage mechanisms to ensure timely processing and validation (Rahman et al., 2023; Xue et al., 2023). Interoperability between blockchain and IoT platforms is another challenge, as the lack of standardized protocols can hinder seamless integration and data exchange. Furthermore, security and privacy are paramount in B-IoT, as the interconnectedness of devices and the potential for unauthorized access pose significant risks (Chen et al., 2023). Ensuring the integrity and confidentiality of data and protecting against potential attacks and vulnerabilities require robust security measures and encryption techniques (Khan et al., 2023). Finally, addressing B-IoT systems' energy consumption and computational requirements is crucial to ensure sustainability and efficiency.

These include the development of lightweight consensus algorithms tailored for IoT devices, integrating edge computing to reduce latency and enhance scalability, and adopting interoperability standards to facilitate seamless communication between blockchain and IoT platforms (Khan et al., 2023; Shafik and Mostafavi, 2019). This research study explores the opportunities, challenges, and potential solutions associated with B-IoT. By conducting a comprehensive literature review, identifying research gaps, proposing methodologies, and analyzing real-world use cases, this chapter seeks to underwrite B-IoT and pave the way for future advancements in this field.

Understanding B-IoT's opportunities and challenges is crucial for unlocking its transformative potential across sectors (Shafik, Matinkhah, and Sanda, 2020). By leveraging the combined power of blockchain and IoT, we can create secure, transparent, and decentralized systems that enhance efficiency, trust, and innovation. Through this research, we aim to shed light on the intricacies of B-IoT, stimulate further exploration, and foster innovation in this exciting domain. Furthermore, regulatory and legal considerations challenge the widespread adoption of B-IoT (Shafik and Matinkhah, 2019b). Implementing blockchain-based systems involves compliance with existing regulations, such as data protection and privacy laws. In addition, the evolving nature of blockchain technology and its intersection with IoT necessitate careful examination of legal frameworks to ensure compliance and address any potential legal hurdles or uncertainties (Shafik, Matinkhah, and Shokoor, 2022).

Security remains a paramount concern in B-IoT deployments. While blockchain technology provides inherent security features, such as immutability and cryptographic integrity, vulnerabilities and attack vectors specific to IoT devices and networks must be addressed (Shafik, Matinkhah, and Ghasemzadeh, 2020). In addition, the diverse range of IoT devices with varying computational capabilities and security measures makes it crucial to develop robust security protocols that can withstand potential attacks, secure device-to-device communication, and protect sensitive data. Various solutions and research efforts are underway (Liu et al., 2023). These include the development of lightweight consensus algorithms optimized for resource-constrained IoT devices, the exploration of hybrid architectures, combining public and private blockchains to balance security and scalability, and the integration of artificial intelligence techniques to enhance anomaly detection and threat mitigation in B-IoT ecosystems (Zhang et al., 2023).

Integrating blockchain technology with the IoT presents significant opportunities to revolutionize industries and domains (Li, Tao, and Gong, 2023; Xu et al., 2023). B-IoT offers enhanced security, transparency, and decentralized data management, enabling novel applications and use cases across sectors. However, challenges such as scalability, interoperability, security, and regulatory compliance must be addressed to unlock the full potential of B-IoT (Qian et al., 2023). Nevertheless, through ongoing research, innovation, and collaboration between academia, industry, and regulatory bodies, the path to harnessing the transformative power of B-IoT can be paved, leading to a more secure, efficient, and trustworthy IoT ecosystem that benefits society as a whole (Qi et al., 2023).

### 3.1.1 KEY CONTRIBUTIONS OF THE CHAPTER

This chapter presents the following significant contributions:

- Portrays the IoT and blockchain technology description and their integration.
- Conducts a comprehensive review of the existing blockchains, IoT, and their convergence literature, and identify the gaps in the existing literature that the current research community needs to address.
- Presents the potential opportunities and advantages of integrating blockchain and IoT.
- Provides examples and use cases where B-IoT can enhance various domains (e.g., healthcare, supply chain, and energy).
- Discusses key challenges and limitations associated with implementing B-IoT solutions; addresses issues such as scalability, interoperability, security, and privacy concerns; and analyzes the implications of these challenges on adopting B-IoT.
- Presents the open research questions and future trends from a B-IoT perspective.

### 3.1.2 ORGANIZATION OF THE CHAPTER

The rest of the chapter is structured into eight sections. Section 3.2 demonstrates the overview and challenges of IoT and blockchain technologies, critical characteristics, and the taxonomy of blockchain systems. Section 3.3 illustrates the integration of the IoT and blockchains, including opportunities to integrate blockchain with IoT. Blockchain of things (BCoT) architecture along with BCoT 6G deployment and beyond are presented. Section 3.4 extensively explores B-IoT and demonstrates the potential opportunities and advantages of integrating blockchain and IoT. It provides examples and use cases where B-IoT can enhance various domains (e.g., healthcare, supply chain, and energy). Section 3.5 provides solutions for B-IoT which include the proposed model and discussion on the potential solutions, frameworks, or architectures to overcome the mentioned challenges. Section 3.6 briefly explores ongoing research efforts or industry initiatives in addressing these challenges. Section 3.7 presents the top open research questions on the IoT and blockchain individually and their integration. Some future directions are depicted. Finally, Section 3.8 offers the lessons learned and the conclusion.

## 3.2 OVERVIEW OF INTERNET OF THINGS AND BLOCKCHAIN TECHNOLOGIES

This section briefly introduces the IoT, including the challenges, critical characteristics of blockchain, and taxonomy of blockchain systems.

IoT has emerged as a transformative concept, garnering considerable attention in recent years. It encompasses a network of interconnected devices, objects, and systems capable of autonomous communication, data exchange, and task execution without human intervention (Zubaydi, Varga, and Molnár, 2023). Initially focused on connecting computers and smartphones, the IoT has evolved to integrate a broad range of everyday objects, leveraging advancements in sensor technology, connectivity, and data processing capabilities (Sharma et al., 2023).

This explosion of connected IoTs has led to an exponential surge in data generation, empowering real-time monitoring, automation, and intelligent decision-making. As a result, the IoT is revolutionizing industries and sectors worldwide (Shafik, Matinkhah, and Ghasemazade, 2019). For example, intelligent cities employ IoT applications to optimize urban services, enhance energy efficiency, and improve public safety. In healthcare, IoT devices enable remote patient monitoring, personalized medicine, and streamlined healthcare delivery (Zhao et al., 2022). Agriculture benefits from IoT-driven precision farming, environmental monitoring, and efficient crop management. Likewise, the industrial sector capitalizes on the IoT for predictive maintenance, supply chain optimization, and process automation (Jun et al., 2021).

These illustrations showcase IoT's diverse applications and transformative impact across sectors. Looking ahead, the prospects of the IoT are vast and promising. With continued technological advancements, the IoT ecosystem is poised to experience exponential growth, connecting billions of devices globally (Shafik, Matinkhah, and Ghasemzadeh, 2019). This immense connectivity will generate colossal volumes of data, fueling advancements in artificial intelligence, machine learning, and data analytics. In addition, edge computing, facilitating data processing and analysis at the network edge, will play a pivotal role in enhancing real-time capabilities and reducing latency in IoT applications (Meng et al., 2020). Furthermore, integrating emerging technologies, for example, edge artificial intelligence, 5G networks, and blockchain, will expand the IoT's capabilities, unlocking new security, scalability, and interoperability possibilities.

### 3.2.1  INTERNET OF THINGS LAYERS

IoT has three main layers: perception, communication, and industrial application, as demonstrated in Figure 3.1.

#### 3.2.1.1  Perception Layer

The perception layer is the foundational layer of the IoT architecture; it comprises physical devices, sensors, and actuators interacting with the physical environment and collecting data. These devices are responsible for perceiving and capturing real-world information such as temperature, humidity, motion, and light (Fotia, Delicato, and Fortino, 2023). The perception layer is critical in empowering the digitization

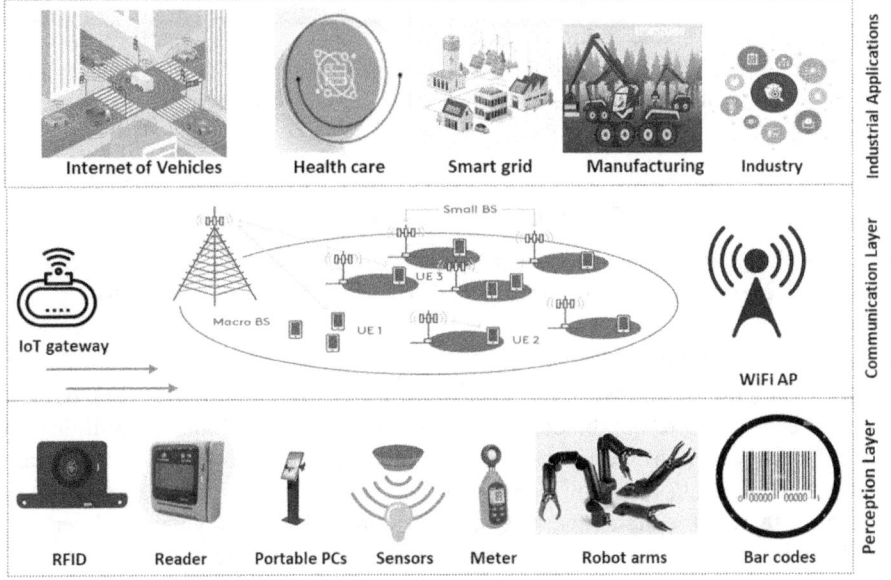

**FIGURE 3.1**    A simplified Internet of Things: layers of application.

of physical objects and environments, as it is the interface between the physical and digital worlds. Furthermore, it facilitates the collection and processing of data, which serves as the basis for subsequent layers of the IoT architecture.

### 3.2.1.2  Communication Layer

The communication layer bridges the perception and higher layers of the IoT architecture. It encompasses various communication technologies and protocols facilitating seamless data exchange between connected devices. In addition, this layer ensures reliable and secure data transmission from the perception layer to the cloud or edge computing platforms for further processing and analysis (Rejeb et al., 2023). Communication technologies in this layer include wireless protocols like Bluetooth, Zigbee, Wi-Fi, and cellular networks like 4G and 5G. The communication layer enables real-time data transfer, interoperability, and connectivity across many IoT devices and networks.

### 3.2.1.3  Industrial Application

The industrial application layer is the topmost layer of the IoT architecture and focuses on leveraging the collected data and insights to drive, value, and deliver specific business outcomes. This layer encompasses applications, platforms, and services that utilize the processed data to enable intelligent decision-making, automation, and optimization in various industrial domains. It involves advanced analytics, machine learning algorithms, and artificial intelligence techniques to derive actionable insights and create intelligent systems (Sánchez et al., 2023). Industrial applications of IoT can be found in sectors such as manufacturing, logistics, energy, healthcare, and agriculture, where IoT-enabled solutions enhance operational efficiency, enable predictive maintenance, optimize resource utilization, and enable data-driven decision-making. The industrial application layer transforms raw data into valuable information, enabling organizations to achieve tangible benefits and gain a competitive edge in their respective industries.

### 3.2.2 Challenges of the Internet of Things

Some inevitable challenges remain persistent as these devices (IoT) connect to the Internet. Therefore, this section presents the top six main challenges of IoT.

#### 3.2.2.1 Security Vulnerability

Security vulnerabilities pose significant challenges in the IoT context. IoT devices' interconnected nature, combined with their varied computational capabilities and limited security measures, creates an environment susceptible to potential attacks and breaches (Zhao et al., 2022). IoT devices often lack robust security features, making them easy targets for malicious actors. Additionally, IoT deployments' sheer scale and diversity make it challenging to ensure uniform security standards and protocols across devices and networks (Jun et al., 2021). Protecting IoT systems from unauthorized access, data breaches, and privacy infringement requires the development of comprehensive security frameworks, encryption techniques, and authentication mechanisms to safeguard the integrity, confidentiality, and availability of IoT data and communications.

#### 3.2.2.2 Privacy Vulnerability

Ensuring privacy in the IoT presents a significant challenge. The proliferation of interconnected devices and the constant generation of personal data raise concerns regarding data privacy and user consent (Sánchez et al., 2023). IoT devices often collect sensitive information, such as personal health or location data, increasing the risk of unauthorized access or misuse. Moreover, the diverse stakeholders involved in IoT ecosystems, including device manufacturers, service providers, and third-party applications, necessitate robust privacy policies and frameworks to protect user privacy throughout the entire data life cycle. Addressing privacy vulnerabilities in IoT requires implementing privacy-by-design principles, data anonymization techniques, and secure data-sharing protocols to establish trust, transparency, and user control over their personal information (Cai et al., 2023).

#### 3.2.2.3 The Complexity of IoT Deployment

In addition to security and privacy vulnerabilities, IoT deployments' sheer scale and complexity present challenges regarding data governance and compliance. The vast amount of data IoT devices generate raises concerns about data ownership, access rights, and regulatory compliance. IoT ecosystems involve multiple stakeholders and data flows, making it essential to establish clear data governance frameworks and address issues such as data sovereignty, data-sharing agreements, and data protection regulations (Mahmood and Al Dabagh, 2023). Ensuring proper data governance in IoT requires transparent data management practices, adherence to data protection regulations, and establishing mechanisms for auditing and enforcing compliance. Effective data governance is critical for building trust, fostering innovation, and maintaining ethical and responsible practices in IoT deployments.

### 3.2.2.4    IoT Resource Constraints

The resource constraints of IoT devices pose a momentous challenge in the IoT landscape. IoT devices often have limited processing power, memory, and energy resources due to their small form factor and battery-powered operation. These constraints restrict the computational capabilities and limit the execution of resource-intensive tasks on IoT devices. For example, a resource-constrained IoT device may struggle to perform complex data analysis or encryption tasks locally (Srivastava et al., 2023). This limitation necessitates offloading computational tasks to edge or cloud computing platforms, which introduces additional latency and dependency on external infrastructure. Mitigating resource constraints in IoT devices requires the development of efficient algorithms, lightweight protocols, and power-efficient designs to optimize resource utilization and enable intelligent processing at the network edge (Chandrakar et al., 2023).

### 3.2.2.5    Poor Interoperability

The lack of standardized communication protocols, data formats, and interoperable systems hinders seamless connectivity and collaboration among IoT devices and platforms. Furthermore, this fragmentation inhibits data exchange and interoperability across different IoT ecosystems, hindering the realization of IoT's full potential (Mannayee and Ramanathan, 2023). For instance, a smart home device may struggle to communicate with a smart city infrastructure due to incompatible protocols. Addressing this challenge requires the development of open standards, protocols, and frameworks that enable interoperability, data harmonization, and seamless integration between disparate IoT systems and devices, fostering a unified and connected IoT ecosystem.

### 3.2.2.6    Heterogeneity

The diverse range of IoT devices, protocols, and platforms with varying capabilities, communication protocols, and data formats results in a complex and heterogeneous ecosystem. This heterogeneity creates interoperability issues, challenging integrating and managing different devices and systems. For example, a smart home device using the Zigbee protocol may face difficulties communicating with an industrial IoT device using the Modbus protocol (Bediya and Kumar, 2023). Overcoming this challenge requires the development of standardized interfaces, protocols, and middleware solutions that enable seamless interoperability and integration across heterogeneous IoT environments, fostering collaboration and enabling the exchange of data and services between various devices and systems.

The challenges of security vulnerabilities, privacy vulnerabilities, resource constraints, poor interoperability, and heterogeneity collectively pose significant obstacles to the widespread adoption and success of the IoT (Han et al., 2023).

Addressing these challenges requires a multi-faceted approach involving the development of robust security and privacy frameworks, optimizing resource utilization, standardization of communication protocols, and harmonizing data formats. Furthermore, fostering collaboration between industry stakeholders, regulatory bodies, and researchers is crucial for overcoming these challenges and ensuring IoT systems' seamless integration, interoperability, and trustworthiness (Issa et al., 2023). By effectively addressing these challenges, the full potential of the IoT can be realized, leading to transformative advancements across various domains.

### 3.2.3  BLOCKCHAIN TECHNOLOGY OVERVIEW

Blockchain technologies have emerged as a groundbreaking innovation with immense potential across various industries. At its core, blockchain is a decentralized and immutable ledger that enables secure and transparent transactions. It eliminates the need for intermediaries by utilizing cryptographic algorithms to verify and record transactions tamper-resistantly (Tyagi et al., 2023). Blockchain's essential features, such as decentralization, transparency, and immutability, provide opportunities for enhanced trust, efficiency, and security in areas like finance, supply chain management, healthcare, and more (Rahman et al., 2023). Various blockchain implementations, including public and private blockchains, smart contracts, and decentralized applications, are being explored and developed to harness the transformative power of this technology.

Furthermore, blockchain technologies offer several advantages over traditional centralized systems. They provide increased security by leveraging consensus algorithms and cryptographic methods to ensure the integrity and immutability of data. The decentralized nature of blockchain networks eliminates single points of failure and reduces the risk of data manipulation or fraud (Xue et al., 2023). Additionally, blockchain enables transparent and auditable transactions, promoting participant trust and accountability. Blockchain technology has recently witnessed significant advancements (Haque et al., 2023). The emergence of new consensus mechanisms, for instance, delegated proof-of-stake and proof-of-stake (PoS), has addressed the scalability and energy efficiency concerns associated with traditional proof-of-work (PoW) systems (Khan et al., 2023).

As demonstrated in Figure 3.2, a transaction is initialized between Alice and Bob; the transaction is passed through a node that broadcasts the transaction to the peer-to-peer network, where the transaction is authenticated. The authenticated transactions are then appended to other transactions to form a block.

In a blockchain, a block is formed through a specific process that involves several key components. When a transaction or a set of transactions is initiated on the blockchain network, it undergoes a series of steps to become part of a block. First, network participants gather and verify the transactions, also known as nodes or miners. Next, these participants validate the integrity and authenticity of the transaction to ensure that they meet the predefined criteria, such as proper digital signatures and available funds (Shafik, 2023). Once the transactions are validated, they are grouped together into a block. The block typically contains a header and a body. The header contains metadata about the block, including a unique identifier (hash) that links it to

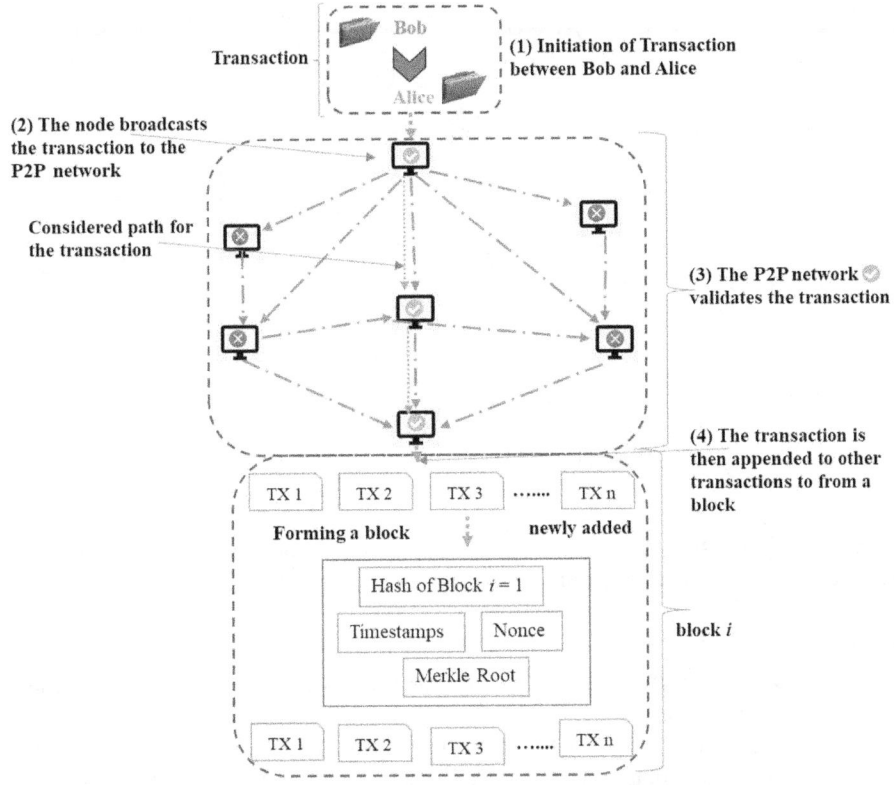

**Transaction**

Bob

Alice

(1) Initiation of Transaction between Bob and Alice

(2) The node broadcasts the transaction to the P2P network

Considered path for the transaction

(3) The P2P network ⊘ validates the transaction

(4) The transaction is then appended to other transactions to from a block

| TX 1 | TX 2 | TX 3 | ······· | TX n |

**Forming a block**                                              newly added

Hash of Block $i = 1$

Timestamps          Nonce

Merkle Root

block $i$

| TX 1 | TX 2 | TX 3 | ······· | TX n |

**FIGURE 3.2**   A blockchain technology illustration.

the previous block, forming a chain. This hash ensures the integrity and immutability of the blockchain by creating a cryptographic link between blocks.

The body of the block contains the actual transactions that have been verified. The transactions are stored in a specific format that varies depending on the blockchain protocol being used. For example, in Bitcoin, transactions are stored in a Merkle tree structure, where multiple transactions are combined and hashed together (Sharma et al., 2023). After the block is formed, it undergoes consensus mechanisms, for example, PoW or PoS, depending on the blockchain protocol. This mechanism ensures that the block is validated by the network participants and added to the existing blockchain securely and trustworthily. Once the block is validated and added to the blockchain, it becomes a permanent part of the ledger (Li et al., 2023; Liu et al., 2023). The new block's hash is used as the reference for the next block in the chain, creating a contiguous sequence of blocks.

A block is formed in a blockchain by gathering, validating, grouping, and linking verified transactions. This ensures the blockchain's integrity, security, and immutability, creating a reliable and transparent ledger of transactions (Liu et al., 2023). An illustration of this is presented in Figure 3.3 with a simple life cycle of technological contracts that involve mainly the stages of creating, deploying, executing,

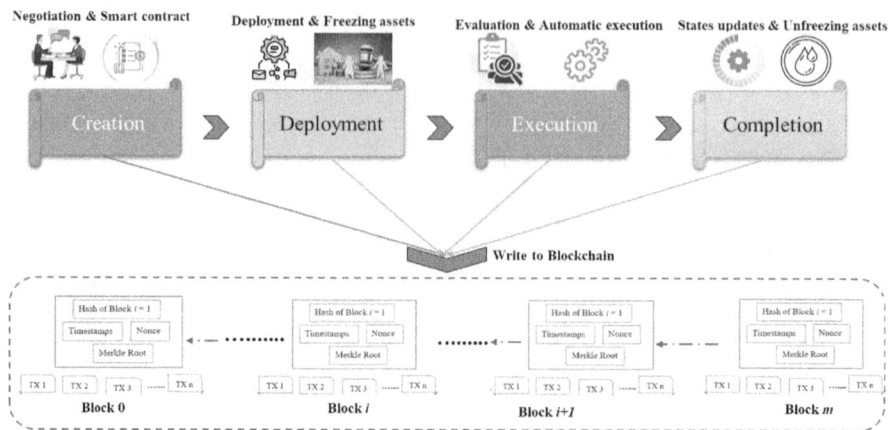

**FIGURE 3.3**  Intelligent contract life cycle.

and completing, where different last three stages are introduced to the blockchain to increase security and privacy during negotiation and contracting at the creation stage.

### 3.2.3.1  Creation

The first phase in the life cycle of a smart contract is its creation. During this phase, the smart contract is designed and developed according to the specific requirements and logic defined by the contract creator. The contract is typically written in a programming language suitable for smart contracts, such as Solidity for Ethereum. The contract creator defines the contract's variables, functions, and conditions that govern its execution. The creation phase considers the contract's purpose, functionality, and potential interactions with other contracts or external systems.

### 3.2.3.2  Deployment

Once the smart contract is created, it enters the deployment phase. The contract is uploaded to the blockchain network and deployed on a specific blockchain platform in this phase. Deployment involves specifying the address and parameters of the contract on the blockchain. The contract is then stored on the blockchain's distributed ledger and becomes accessible to network participants. During deployment, the contract creator may also define initial values for the contract's variables or configure any necessary permissions or access controls.

### 3.2.3.3  Execution

After deployment, the intelligent contract enters the execution phase. This is when authorized parties can invoke and execute the contract's functions on the blockchain network. The execution phase involves interacting with the contract through transactions, which trigger the execution of specific functions within the contract. These functions may involve modifying the contract's state, retrieving data, or performing complex operations based on the contract's predefined logic. In addition, the execution phase relies on the consensus mechanism of the blockchain network to validate and record the contract's transactions and ensure their integrity.

### 3.2.3.4  Completion

The final phase in the life cycle of a smart contract is completion. This phase occurs when the contract has fulfilled its intended purpose or reaches a specified condition for termination. Completion can be triggered by the contract's predefined logic, external events, or manual intervention. Upon completion, the contract may perform final operations such as transferring assets, updating states, or emitting events to notify other participants. Therefore, it is essential to adequately handle contract completion to ensure any necessary cleanup, closure of resources, or termination of ongoing processes.

## 3.3  CONVERGENCE OF BLOCKCHAIN AND IOT

Integrating IoT and blockchain technologies holds significant promise in addressing the challenges of security, privacy, data integrity, and interoperability in IoT deployments. By combining blockchain's decentralized and secure nature with the extensive connectivity of IoT devices, new opportunities for enhanced trust, transparency, and efficiency arise (Zhang et al., 2023). Blockchain's distributed ledger and cryptographic algorithms can secure IoT device identity, data transmission, and access control mechanisms (Liu et al., 2023). Furthermore, through smart contracts, blockchain can establish automated and tamper-proof trust among IoT devices, ensuring the authenticity and integrity of data.

Furthermore, blockchain can address data privacy concerns in IoT. Using encryption and decentralized data storage, blockchain can give users greater control over their data and enable selective data sharing. In addition, IoT devices can authenticate and authorize data access using blockchain-based identity management systems, enhancing privacy and consent management. Another advantage of integrating IoT and blockchain is data integrity and audibility (Mannayee and Ramanathan, 2023). Blockchain's immutability ensures that IoT data recorded on the blockchain remains tamper-proof and transparent. This can be particularly beneficial in applications such as supply chain management, where tracing the origin and movement of goods can be reliably documented on the blockchain, enhancing transparency and accountability (Yang et al. 2021).

Interoperability between IoT devices and platforms is another challenge that blockchain can address. Blockchain-based protocols and standards can facilitate seamless communication, data exchange, and interoperability among IoT devices and networks. Smart contracts can enable automated and self-executing transactions between IoT devices, eliminating the need for intermediaries and enhancing the efficiency of IoT ecosystems (Shafik, Matinkhah, and Ghasemazade, 2019). However, integrating IoT and blockchain also presents particular challenges. The resource restraints of IoT, like limited processing power, may hinder direct integration with resource-intensive blockchain networks. In addition, scalability and latency issues must be carefully addressed to handle the large-scale data generated by IoT devices (Matinkhah and Shafik, 2019; Reshi and Sholla, 2022).

Additionally, the complexity of managing blockchain infrastructure and ensuring consensus among IoT devices pose technical challenges that must be tackled. By leveraging both technologies' strengths, innovative applications can be developed in supply chain management, healthcare, energy, and other sectors (Shafik and Matinkhah, 2019a). However, careful consideration of technical challenges and the efficient utilization of resources are essential to unlock the full potential of this integration and drive the widespread adoption of secure and trustworthy IoT systems.

## 3.4  BLOCKCHAIN INTEGRATION WITH IOT OPPORTUNITIES

In this section, we present some trending opportunities, such as integrating blockchain technology and IoT.

### 3.4.1  ENHANCED SECURITY

Blockchain can provide a robust security framework for IoT devices by leveraging its decentralized and tamper-resistant nature. It enables secure and transparent transactions, authentication, and data encryption, reducing the risk of unauthorized access, data manipulation, and cyberattacks. Integrating a blockchain with IoT enhances security by leveraging its decentralized and tamper-resistant nature (Haque et al., 2023; Zhao et al., 2022). A blockchain-based platform uses its Tangle technology to provide secure data transfer and storage for IoT devices. It ensures data integrity and prevents unauthorized access, enabling secure communication and transactions. Another example is Filament, which combines blockchain with hardware security modules to create a secure network for IoT devices, protecting against tampering and unauthorized device access. These examples demonstrate how blockchain enhances the security of IoT devices, safeguarding sensitive data and mitigating cybersecurity risks.

### 3.4.2  IMPROVED DATA INTEGRITY

Integrating a blockchain with IoT enhances security by leveraging its decentralized and tamper-resistant nature. As a result, it ensures data integrity and prevents unauthorized access, enabling secure communication and transactions. Another example is Filament, which combines blockchain with hardware security modules to create a secure network for IoT devices, protecting against tampering and unauthorized device access (Singh, Sturley, and Tewari, 2023). These examples demonstrate how blockchain enhances the security of IoT devices, safeguarding sensitive data and mitigating cybersecurity risks.

### 3.4.3  EFFICIENT DEVICE MANAGEMENT

Blockchain simplifies device management in IoT ecosystems. For example, Chronicled employs blockchain to create an identity registry for IoT devices, enabling secure device onboarding and authentication. This streamlines device management processes, reducing administrative overhead and enhancing operational efficiency (Mannayee and Ramanathan, 2023). Similarly, IOTA's blockchain-based approach allows IoT devices to manage their identities and transactions autonomously, enabling self-sovereign device management. These examples showcase blockchain streamlines, device management, enabling seamless integration, and secure communication among IoT devices (Fotia et al., 2023).

### 3.4.4 STREAMLINED TRANSACTIONS AND PAYMENTS

Blockchain technology enables seamless and secure peer-to-peer transactions and micropayments between IoT devices, eliminating the need for intermediaries. This can simplify and automate payment processes within IoT ecosystems, enabling machine-to-machine transactions and new business models. In addition, blockchain facilitates trusted data sharing and monetization in IoT networks. For instance, Streamr enables real-time data monetization through its blockchain-based data marketplace (Singh et al., 2023). It enables IoT devices to securely share and sell data, empowering individuals and organizations to leverage the value of their data. Another example is Ocean Protocol, which utilizes blockchain to create a decentralized data exchange where IoT data providers can share and monetize their data. These examples demonstrate how blockchain fosters trusted data sharing and incentivizes monetization, creating new opportunities in the IoT ecosystem (Chandrakar et al., 2023).

### 3.4.5 SCALABILITY AND INTEROPERABILITY

Integrating blockchain with IoT can address scalability and interoperability challenges by providing standardized protocols, data formats, and decentralized consensus mechanisms. It enables seamless integration and communication between various IoT devices, platforms, and networks, facilitating interoperability and expanding the potential applications of IoT (Chandrakar et al., 2023). This enables new business models, such as pay-per-use or peer-to-peer energy trading. Similarly, Waltonchain combines blockchain and IoT for retail applications, enabling automatic payment and inventory management through radio-frequency identification technology (Sánchez et al., 2023). These examples highlight how blockchain streamlines transactions and payments in IoT, enabling efficient and frictionless value exchange between devices.

## 3.5 BLOCKCHAIN OF THINGS ARCHITECTURE

Within this research, based on the fundamental layers of IoT (Figure 3.1), we propose a blockchain composite layer between the communication layer and industrial application (Figure 3.4). Figure 3.5 demonstrates the section at which this composite layer is situated.

Within the communication and manufacturing application layers of the IoT, composite sublayers contribute to the overall functionality and operation of the IoT

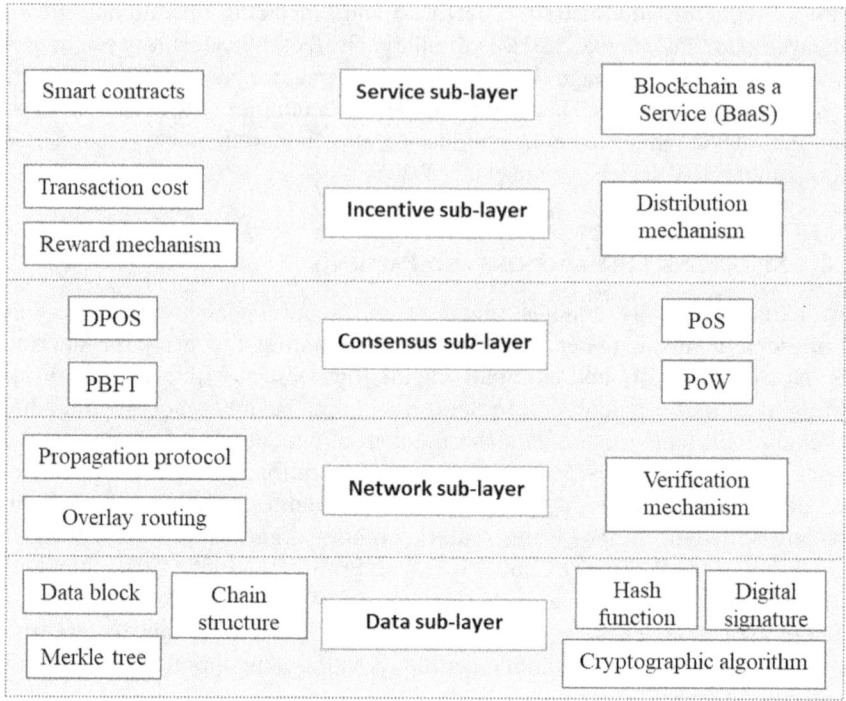

**FIGURE 3.4**  A proposed composite layer entails data, network, consensus, incentive, and service sublayers.

ecosystem of blockchain assisted systems. The composite layer has five sublayers, each having a specific function in operation, as demonstrated briefly below.

### 3.5.1  DATA SUBLAYER

The data sublayer collects, stores, and manages data generated by IoT devices. It encompasses data acquisition, preprocessing, aggregation, and storage techniques. In addition, this sublayer focuses on ensuring the quality, security, and integrity of IoT data and enabling efficient data retrieval and analysis for decision-making processes.

### 3.5.2  NETWORK SUBLAYER

The network sublayer involves establishing and managing communication networks that interconnect IoT devices. It encompasses protocols, network architectures, and technologies for enabling seamless and reliable data exchange between devices. In addition, this sublayer addresses connectivity, bandwidth, latency, and scalability challenges to support the vast number of interconnected IoT devices.

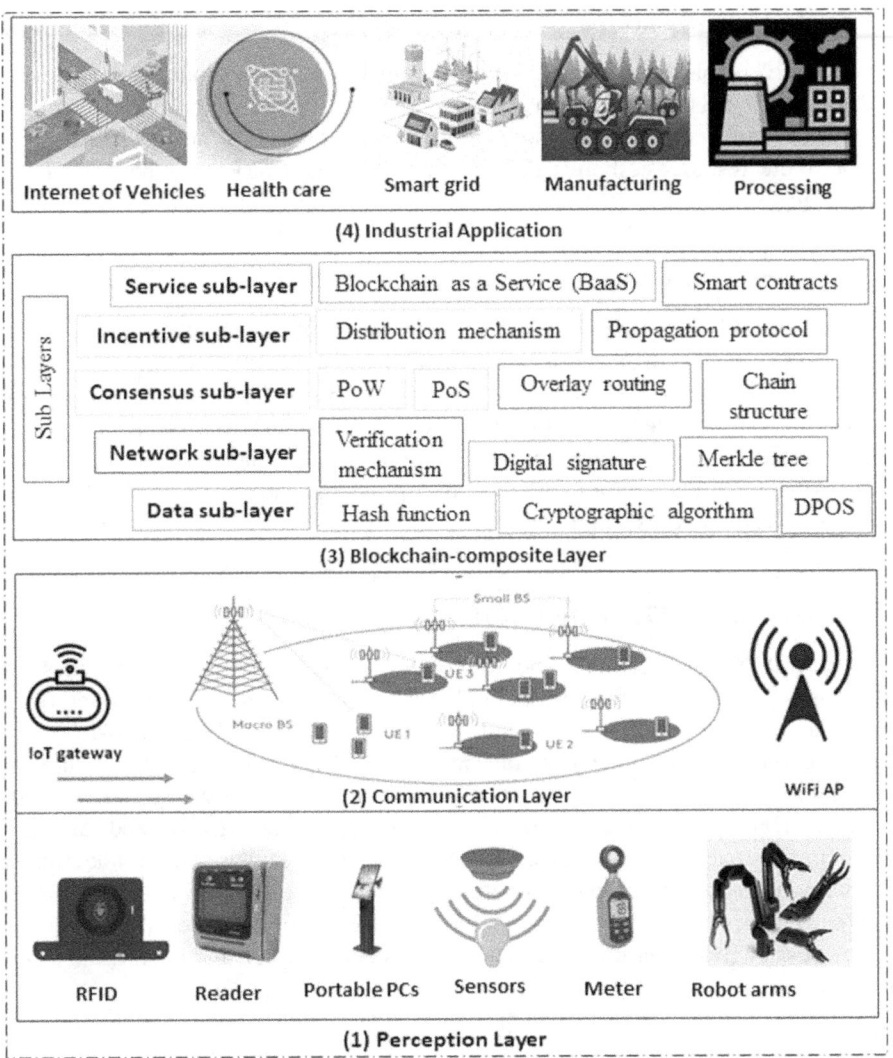

**FIGURE 3.5** An integrated composited blockchain technology and the Internet of Things.

### 3.5.3 Consensus Sublayer

The consensus sublayer is responsible for achieving agreement and consensus among the nodes or participants in the IoT network. It utilizes consensus algorithms and mechanisms to ensure that the shared data and state of the IoT system are consistent across all participants. This sublayer plays a crucial role in maintaining the integrity and trustworthiness of the distributed ledger or blockchain that may be utilized within the IoT ecosystem.

### 3.5.4 Incentive Sublayer

The incentive sublayer provides incentives and rewards to encourage participation and cooperation within the IoT network. It leverages tokenization, smart contracts, or other incentive structures to motivate IoT device owners or network participants to contribute resources, share data, or perform specific tasks that benefit the IoT ecosystem.

### 3.5.5 Service Sublayer

The service sublayer encompasses the provision of value-added services and applications that leverage the data and capabilities of the IoT devices. It involves developing, deploying, and managing IoT applications that cater to specific industry requirements or user needs. This sublayer includes functionalities such as analytics, visualization, control interfaces, and integration with external systems or platforms. They form the foundation for the successful deployment and operation of IoT systems across various industries and domains.

## 3.6 OPPORTUNITIES OF B-IOT

The integration of blockchain and IoT offers several potential opportunities and advantages. This section provides examples and use cases where the proposed B-IoT can enhance various domains like healthcare, supply chain, and energy. B-IoT offers improved security, data integrity, and decentralized control. It enables secure and transparent data exchange, authentication, and tamper-resistant storage, reducing the risk of data breaches and unauthorized access. B-IoT facilitates trusted and automated transactions between IoT devices, enabling new business models and revenue streams. Furthermore, B-IoT fosters data monetization, incentivizes data sharing, and enables secure peer-to-peer communication and cooperation among devices. Apart from the deployment of BCoT, various interaction modes between IoT and blockchain can be implemented.

### 3.6.1 Solutions for B-IoT and 5G and Beyond

These solutions encompass technical advancements, standardization efforts, and collaborative approaches, as demonstrated below.

#### 3.6.1.1 Scalable and Efficient Blockchain Protocols

Developing scalable and efficient blockchain protocols specifically designed for IoT environments is crucial. These protocols should consider the resource-constrained nature of IoT devices, optimizing consensus mechanisms, data storage, and transaction processing to minimize computational and storage requirements (Srivastava et al., 2023). Examples include lightweight consensus algorithms like directed acyclic graphs or proof-of-authority structures like IOTA's Tangle, which offer scalability and efficiency suitable for IoT deployments (Xu et al., 2023).

### 3.6.1.2 Privacy-Preserving Mechanisms

Privacy in B-IoT systems is essential to protect sensitive data from IoT devices. Solutions like zero-knowledge proof, homomorphic encryption, and differential privacy techniques can be employed to ensure data confidentiality while enabling data sharing and analysis (Li et al., 2023). By implementing privacy-enhancing technologies, B-IoT systems can strike a balance between data privacy and the transparency provided by blockchain technology.

### 3.6.1.3 Interoperability Standards and Frameworks

Establishing interoperability standards and frameworks is crucial for seamless integration and collaboration among IoT devices, blockchain platforms, and applications. Efforts such as developing open-source interoperability protocols like Hyperledger or adopting industry standards can facilitate cross-platform compatibility and enable the exchange of data and services in a secure and standardized manner (Liu et al., 2023). Collaborative approaches involving industry consortia, research communities, and regulatory bodies are also necessary to address the challenges and develop comprehensive solutions for B-IoT (Qi et al., 2023). These solutions must focus on scalability, privacy, and interoperability to unlock the full potential of integrating blockchain with IoT, fostering innovation, and driving the adoption of B-IoT across various industries and applications, including smart manufacturing, supply chain, food industry, Internet of Vehicles and unmanned aerial vehicle, healthcare, and smart grid, among others, using the functional framework demonstrated in Figure 3.6.

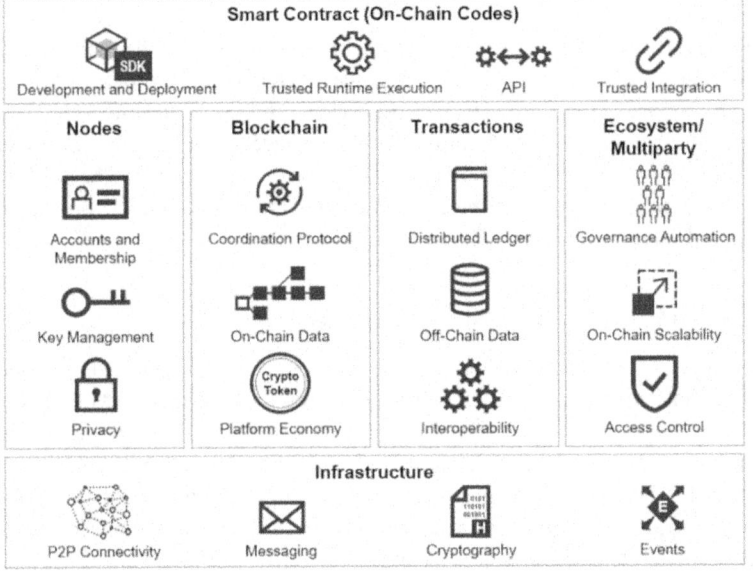

**FIGURE 3.6** 5G beyond blockchain functional framework.

## 3.7   OPEN RESEARCH QUESTIONS AND FUTURE DIRECTIONS

Integrating blockchain with IoT, known as B-IoT, presents exciting opportunities, but it also raises several open research questions and offers directions for future exploration. These focus areas can contribute to further advancements and improvements in B-IoT systems. Some key research questions and future directions are presented here.

### 3.7.1   SCALABILITY AND PERFORMANCE OPTIMIZATION

One of the primary challenges in B-IoT is achieving scalability and optimizing the performance of blockchain protocols in IoT environments. Future research can explore innovative consensus algorithms, data partitioning techniques, and sharing approaches tailored explicitly for resource-constrained IoT devices. In addition, finding ways to enhance transaction throughput, reduce latency, and minimize energy consumption will enable large-scale deployment of B-IoT systems.

### 3.7.2   PRIVACY AND SECURITY ENHANCEMENTS

As privacy and security remain significant concerns in IoT, future research should focus on developing robust privacy-preserving mechanisms and enhancing security protocols for B-IoT. For example, exploring techniques like zero-knowledge proof, secure multiparty computation, and trusted execution environments can ensure data confidentiality, integrity, and access control in decentralized IoT environments. In addition, investigating potential vulnerabilities, attack vectors, and countermeasures specific to B-IoT can help mitigate security risks.

### 3.7.3   STANDARDS AND INTEROPERABILITY

Establishing standards and interoperability frameworks is vital to facilitate seamless communication and collaboration among various IoT devices, blockchain platforms, and applications. Therefore, future research can concentrate on developing standardized protocols, data formats, and interfaces that enable interoperability between different B-IoT solutions. Furthermore, exploring methods for cross-blockchain interoperability and integration with existing IoT standards will foster the adoption and integration of B-IoT in real-world scenarios.

### 3.7.4   GOVERNANCE AND LEGAL FRAMEWORKS

The decentralized nature of blockchain and IoT raises governance and legal challenges that must be addressed. Future research can focus on developing governance models, consensus on decision-making processes, and regulatory frameworks to ensure accountability, transparency, and compliance within B-IoT ecosystems. Understanding the legal implications, data ownership, liability, and dispute resolution mechanisms in decentralized and autonomous IoT systems is crucial for the sustainable and responsible deployment of B-IoT.

### 3.7.5  REAL-WORLD APPLICATIONS AND USE CASES

Exploring and evaluating practical use cases and applications of B-IoT across various domains is an important area of future research. For example, investigating the potential impact of B-IoT in industries such as supply chain management, healthcare, energy, and smart cities can provide insights into the benefits, challenges, and implementation considerations specific to different contexts. Furthermore, conducting pilot projects and real-world deployments will help validate the feasibility, scalability, and economic viability of B-IoT solutions. By addressing these open research questions and exploring future directions, researchers can contribute to developing robust, scalable, and secure B-IoT systems that unlock blockchain technology's full potential in the IoT.

## 3.8  LESSONS LEARNED AND CONCLUSION

This section presents some lessons learned in the review and the conclusion of the study.

### 3.8.1  LESSONS LEARNED

B-IoT integration holds immense opportunities for industries and domains. By leveraging blockchain technology's security, transparency, and decentralization, B-IoT offers enhanced data integrity, trusted transactions, efficient device management, and scalable applications. We briefly present some lessons learned from the chapter.

#### 3.8.1.1  Scalability Challenges

According to the reviews, one of the primary obstacles in B-IoT is the scalability of blockchain networks. As the number of connected devices increases and the volume of transactions grows, traditional blockchain architectures face limitations in processing capacity. To overcome this challenge, researchers and developers are exploring innovative solutions such as sharding, which involves partitioning the blockchain into smaller parts to improve performance. Other approaches include sidechains, which enable parallel processing of transactions, and off-chain protocols that reduce the on-chain burden.

#### 3.8.1.2  Security Considerations in B-IoT

Security is a critical concern in B-IoT due to the exchange of sensitive data and commands between devices. Blockchain technology offers inherent security features such as cryptographic techniques and distributed consensus. However, additional measures such as encryption, digital signatures, and robust identity management are essential to protect against cyber threats and unauthorized access. Secure essential management practices and access control mechanisms are crucial to maintaining data integrity and confidentiality in B-IoT systems.

### 3.8.1.3   Interoperability and Explainability

B-IoT involves diverse devices and blockchain networks, necessitating seamless integration, interoperability, and total explainability. Standardization efforts are crucial to define standard protocols, data formats, and communication mechanisms, enabling devices from different manufacturers and blockchain platforms to interoperate effectively. Developing interoperability frameworks and open application programming interfaces can facilitate data exchange, enhance collaboration, and drive widespread adoption of B-IoT solutions.

### 3.8.1.4   Energy Efficiency Challenges in B-IoT Deployments

It has been revealed that limited energy sources often power IoT devices, while blockchain protocols can be computationally intensive and energy-consuming. Achieving energy efficiency in B-IoT systems is crucial to ensure long battery life and sustainable operation. Future research should focus on optimizing consensus mechanisms, exploring lightweight blockchain protocols, and implementing energy-saving strategies, such as selective validation and offloading computations to more powerful devices.

### 3.8.1.5   Diverse Opportunities across Industries

B-IoT presents immense opportunities across various industries. In supply chain management, blockchain enables end-to-end traceability, transparency, and accountability, reducing fraud and counterfeit products. B-IoT enables the secure sharing of medical records, efficient drug supply chain management, and remote patient monitoring in healthcare. Energy sector can benefit from B-IoT by facilitating peer-to-peer energy trading and optimizing energy consumption. Smart cities can leverage B-IoT for efficient resource management, traffic control, and environmental monitoring. These opportunities lead to streamlined operations, cost savings, improved data integrity, and enhanced automation.

### 3.8.1.6   Privacy Preservation Challenges in B-IoT

Preserving privacy becomes crucial with the proliferation of IoT devices and the potential for extensive data collection. B-IoT systems must employ privacy-enhancing techniques to safeguard user data. Zero-knowledge proofs, homomorphic encryption, and decentralized identity management systems can enable selective data disclosure and minimize the exposure of sensitive information. Therefore, it is learned that balancing data utility and privacy protection is vital to building trust and ensuring user acceptance of B-IoT solutions.

### 3.8.1.7   Governance Models and Regulatory Frameworks

As B-IoT becomes more prevalent, establishing governance models and regulatory frameworks is essential. These frameworks guide ethical practices, ensure accountability, and address data protection, cybersecurity, and consumer rights concerns. Collaboration between industry stakeholders, policymakers, and researchers is crucial to shape these frameworks and promote responsible deployment of B-IoT technologies.

### 3.8.1.8   Future Trends in B-IoT

Several trends are expected to shape the future of B-IoT. Hybrid blockchain architectures, which combine the strengths of public and private blockchains, offer enhanced scalability, privacy, and flexibility. Integrating B-IoT with edge computing enables real-time processing and reduces latency by performing computations closer to the data source. Artificial intelligence and machine learning techniques can be leveraged to analyze vast volumes of IoT data, extract meaningful insights, and enable intelligent decision-making in B-IoT systems. Moreover, adopting new consensus algorithms tailored for IoT environments, such as proof-of-stake or practical Byzantine fault tolerance, can improve the efficiency and scalability of B-IoT networks.

## 3.8.2   Conclusion

The emergence of B-IoT has brought about significant advancements, and also posed several challenges, in the realm of connecting physical devices to the digital world. This chapter has explored various aspects of B-IoT, including its challenges, solutions, opportunities, open research questions, and future trends. The challenges faced by B-IoT include scalability, security, interoperability, and energy efficiency, among others. However, researchers and industry experts have proposed several solutions to address these challenges, such as consensus mechanisms, encryption techniques, standardization efforts, and energy optimization strategies. Furthermore, B-IoT offers numerous opportunities in diverse domains, such as supply chain management, healthcare, energy, and smart cities, enabling transparent and efficient transactions, improved data integrity, and enhanced automation. Despite the progress made, several open research questions in B-IoT remain, including privacy preservation, governance models, incentive mechanisms, and regulatory frameworks. These questions necessitate further exploration and innovation from academia, industry, and policymakers. The future trends of B-IoT encompass advancements in hybrid blockchain architectures, edge computing integration, artificial intelligence and machine learning applications, and the development of new consensus algorithms tailored for IoT environments. Moreover, adopting B-IoT will likely accelerate as the technology matures, interoperability standards are established, and trust in decentralized systems increases. As B-IoT continues to evolve, fostering collaboration between researchers, practitioners, and policymakers is essential to address the challenges, seize opportunities, and find solutions to the open research questions. Finally, B-IoT has the potential to revolutionize various industries, enable new business models, and enhance the overall functionality and security of IoT ecosystems, paving the way for a more connected and decentralized future.

## REFERENCES

Bediya, Arun Kumar, and Rajendra Kumar. 2023. "A Novel Intrusion Detection System for Internet of Things Network Security." *Research Anthology on Convergence of Blockchain, Internet of Things, and Security*, 330–48. IGI Global.

Cai, Ting, Yuxin Wu, Hui Lin, and Yu Cai. 2023. "Blockchain-Empowered Big Data Sharing for Internet of Things." *Research Anthology on Convergence of Blockchain, Internet of Things, and Security*, 278–90. IGI Global.

Chandrakar, Palak, Rashi Bagga, Yogesh Kumar, Sanjeev Kumar Dwivedi, and Ruhul Amin. 2023. "Blockchain Based Security Protocol for Device to Device Secure Communication in Internet of Things Networks." *Security and Privacy* 6(1):e267.

Chen, Xiaohong, Caicai He, Yan Chen, and Zhiyuan Xie. 2023. "Internet of Things (IoT)—Blockchain-Enabled Pharmaceutical Supply Chain Resilience in the Post-Pandemic Era." *Frontiers of Engineering Management* 10(1):82–95.

Fotia, Lidia, Flávia Delicato, and Giancarlo Fortino. 2023. "Trust in Edge-Based Internet of Things Architectures: State of the Art and Research Challenges." *ACM Computing Surveys* 55(9):1–34.

Han, Pengchong, Zhouyang Zhang, Shan Ji, Xiaowan Wang, Liang Liu, and Yongjun Ren. 2023. "Access Control Mechanism for the Internet of Things Based on Blockchain and Inner Product Encryption." *Journal of Information Security and Applications* 74:103446.

Haque, Md Alimul, Shameemul Haque, Sana Zeba, Kailash Kumar, Sultan Ahmad, Moidur Rahman, Senapathy Marisennayya, and Laiq Ahmed. 2023. "Sustainable and Efficient E-Learning Internet of Things System through Blockchain Technology." *E-Learning and Digital Media*. https://doi.org/10.1177/20427530231156711

Huang, Ruihang, Xiaoming Yang, and P. Ajay. 2023. "Consensus Mechanism for Software-Defined Blockchain in Internet of Things." *Internet of Things and Cyber-Physical Systems* 3:52–60.

Issa, Wael, Nour Moustafa, Benjamin Turnbull, Nasrin Sohrabi, and Zahir Tari. 2023. "Blockchain-Based Federated Learning for Securing Internet of Things: A Comprehensive Survey." *ACM Computing Surveys* 55(9):1–43.

Jun, Yao, Alisa Craig, Wasswa Shafik, and Lule Sharif. 2021. "Artificial Intelligence Application in Cybersecurity and Cyberdefense." *Wireless Communications and Mobile Computing* 2021:1–10.

Khan, Abdullah Ayub, Asif Ali Laghari, Peng Li, Mazhar Ali Dootio, and Shahid Karim. 2023. "The Collaborative Role of Blockchain, Artificial Intelligence, and Industrial Internet of Things in Digitalization of Small and Medium-Size Enterprises." *Scientific Reports* 13(1):1656.

Li, Yihong, Qi Tao, and Yadong Gong. 2023. "Digital Twin Simulation for Integration of Blockchain and Internet of Things for Optimal Smart Management of PV-Based Connected Microgrids." *Solar Energy* 251:306–14.

Liu, Yijia, Jie Wang, Zheng Yan, Zhiguo Wan, and Riku Jäntti. 2023. "A Survey on Blockchain-Based Trust Management for Internet of Things." *IEEE Internet of Things Journal* 10(7): 5898–5922.

Mahmood, Mahmood Subhy, and Najla Badie Al Dabagh. 2023. "Blockchain Technology and Internet of Things: Review, Challenge and Security Concern." *International Journal of Electrical and Computer Engineering* 13(1):718.

Mannayee, Vivekraj, and Thirumalai Ramanathan. 2023. "An Efficient SDFRM Security System for Blockchain Based Internet of Things." *Intelligent Automation & Soft Computing* 35(2): 1545–1563.

Matinkhah, S. Moitaba, and Wasswa Shafik. 2019. "Smart Grid Empowered by 5G Technology." *2019 Smart Grid Conference (SGC)*, 1–6. IEEE.

Meng, Hao, S. Mojtaba Matinkhah, Wasswa Shafik, and Zubair Ahmad. 2020. "A 5G Beam Selection Machine Learning Algorithm for Unmanned Aerial Vehicle Applications." *Wireless Communications and Mobile Computing* 2020:1–16.

Qi, Liuling, Junfeng Tian, Mengjia Chai, and Hongyun Cai. 2023. "LightPoW: A Trust Based Time-Constrained PoW for Blockchain in Internet of Things." *Computer Networks* 220:109480.

Qian, Kai, Yinqiu Liu, Xiaoming He, Miao Du, Suofei Zhang, and Kun Wang. 2023. "HPCchain: A Consortium Blockchain System Based on CPU-FPGA Hybrid PUF for Industrial Internet of Things." *IEEE Transactions on Industrial Informatics* 19(11): 11205–11215.

Rahman, Anichur, Jahidul Islam, Dipanjali Kundu, Razaul Karim, Ziaur Rahman, Shahab S. Band, Mehdi Sookhak, Prayag Tiwari, and Neeraj Kumar. 2023. "Impacts of Blockchain in Software-Defined Internet of Things Ecosystem with Network Function Virtualization for Smart Applications: Present Perspectives and Future Directions." *International Journal of Communication Systems* e5429.

Rejeb, Abderahman, Karim Rejeb, Horst Treiblmaier, Andrea Appolloni, Salem Alghamdi, Yaser Alhasawi, and Mohammad Iranmanesh. 2023. "The Internet of Things (IoT) in Healthcare: Taking Stock and Moving Forward." *Internet of Things* 22: 100721.

Reshi, Iraq Ahmad, and Sahil Sholla. 2022. "Challenges for Security in IoT, Emerging Solutions, and Research Directions." *International Journal of Computing and Digital Systems*, 12(1):1231–1241.

Sánchez, Luis, Jorge Lanza, Iván González, Juan Ramón Santana, and Pablo Sotres. 2023. *Development of a Blockchain-Based Marketplace for the Internet of Things Data*. ICT Express.

Sasikumar, A., Subramaniyaswamy Vairavasundaram, Ketan Kotecha, V. Indragandhi, Logesh Ravi, Ganeshsree Selvachandran, and Ajith Abraham. 2023. "Blockchain-Based Trust Mechanism for Digital Twin Empowered Industrial Internet of Things." *Future Generation Computer Systems* 141:16–27.

Shafik, Wasswa. 2023. "A Comprehensive Cybersecurity Framework for Present and Future Global Information Technology Organizations." In *Effective Cybersecurity Operations for Enterprise-Wide Systems*, 56–79. IGI Global.

Shafik, Wasswa, Mojtaba Matinkhah, Melika Asadi, Zahra Ahmadi, and Zahra Hadiyan. 2020. "A Study on Internet of Things Performance Evaluation." *Journal of Communications Technology, Electronics and Computer Science* 2020:1–19.

Shafik, Wasswa, Mojtaba Matinkhah, and Mamman Nur Sanda. 2020. "Network Resource Management Drives Machine Learning: A Survey and Future Research Direction." *Journal of Communications Technology, Electronics and Computer Science* 2020:1–15.

Shafik, Wasswa, and S. Mojtaba Matinkhah. 2019a. "Admitting New Requests in Fog Networks According to Erlang B Distribution." *2019 27th Iranian Conference on Electrical Engineering (ICEE),* 2016–21. IEEE.

Shafik, Wasswa, and S. Mojtaba Matinkhah. 2019b. "Privacy Issues in Social Web of Things." *2019 5th International Conference on Web Research (ICWR),* 208–14. IEEE.

Shafik, Wasswa, S. Mojtaba Matinkhah, and Mohammad Ghasemazade. 2019a. "Fog-Mobile Edge Performance Evaluation and Analysis on Internet of Things." *Journal of Advance Research in Mobile Computing* 1(3):1–17.

Shafik, Wasswa, S. Mojtaba Matinkhah, and Mohammad Ghasemzadeh. 2019b. "A Fast Machine Learning for 5g Beam Selection for Unmanned Aerial Vehicle Applications." *Information Systems & Telecommunication* 7(28):262–78.

Shafik, Wasswa, S. Mojtaba Matinkhah, and Mohammad Ghasemzadeh. 2020. "Theoretical Understanding of Deep Learning in UAV Biomedical Engineering Technologies Analysis." *SN Computer Science* 1:1–13.

Shafik, Wasswa, S. Mojtaba Matinkhah, Mamman Nur Sanda, and Fawad Shokoor. 2021. "Internet of Things-Based Energy Efficiency Optimization Model in Fog Smart Cities." *JOIV: International Journal on Informatics Visualization* 5(2):105–12.

Shafik, Wasswa, S. Mojtaba Matinkhah, and Fawad Shokoor. 2022. "Recommendation System Comparative Analysis: Internet of Things Aided Networks." *EAI Endorsed Transactions on Internet of Things* 8(29): 1–16.

Shafik, Wasswa, and Seyed Akabr Mostafavi. 2019. "Knowledge Engineering on Internet of Things through Reinforcement Learning." *International Journal of Computer Applications* 177(44):0975–8887.

Sharma, Prakash Chandra, Md Rashid Mahmood, Hiral Raja, Narendra Singh Yadav, Brij B. Gupta, and Varsha Arya. 2023. "Secure Authentication and Privacy-Preserving Blockchain for Industrial Internet of Things." *Computers and Electrical Engineering* 108:108703.

Singh, Raman, Sean Sturley, and Hitesh Tewari. 2023. "Blockchain-Enabled Chebyshev Polynomial-Based Group Authentication for Secure Communication in an Internet of Things Network." *Future Internet* 15(3):96.

Srivastava, Sanskar, Rohit Bansal, Gulshan Soni, and Amit Kumar Tyagi. 2023. "Blockchain Enabled Internet of Things: Current Scenario and Open Challenges for Future." *Innovations in Bio-Inspired Computing and Applications: Proceedings of the 13th International Conference on Innovations in Bio-Inspired Computing and Applications (IBICA 2022) Held During December 15–17, 2022.* Springer, 640–48.

Tyagi, Amit Kumar, Sathian Dananjayan, Deepshikha Agarwal, and Hasmath Farhana Thariq Ahmed. 2023. "Blockchain—Internet of Things Applications: Opportunities and Challenges for Industry 4.0 and Society 5.0." *Sensors* 23(2):947.

Xu, Jinying, Jinfeng Lou, Weisheng Lu, Liupengfei Wu, and Chen Chen. 2023. "Ensuring Construction Material Provenance Using Internet of Things and Blockchain: Learning from the Food Industry." *Journal of Industrial Information Integration* 33:100455.

Xue, He, Dajiang Chen, Ning Zhang, Hong-Ning Dai, and Keping Yu. 2023. "Integration of Blockchain and Edge Computing in Internet of Things: A Survey." *Future Generation Computer Systems* 144:307–26.

Yang, Zhai, Liu Jianjun, Humaira Faqiri, Wasswa Shafik, Alanazi Talal Abdulrahman, M. Yusuf, and A. M. Sharawy. 2021. "Green Internet of Things and Big Data Application in Smart Cities Development." *Complexity* 2021:1–15.

Zhang, Lihua, Boping Li, Haodong Fang, Ganzhe Zhang, and Chunhui Liu. 2023. "An Internet of Things Access Control Scheme Based on Permissioned Blockchain and Edge Computing." *Applied Sciences* 13(7):4167.

Zhao, Liguo, Derong Zhu, Wasswa Shafik, S. Mojtaba Matinkhah, Zubair Ahmad, Lule Sharif, and Alisa Craig. 2022. "Artificial Intelligence Analysis in Cyber Domain: A Review." *International Journal of Distributed Sensor Networks* 18(4):15501329221084882.

Zubaydi, Haider Dhia, Pál Varga, and Sándor Molnár. 2023. "Leveraging Blockchain Technology for Ensuring Security and Privacy Aspects in Internet of Things: A Systematic Literature Review." *Sensors* 23(2):788.

# 4 Revolutionizing IoT with Blockchain

## A State-of-the-Art Review

Saba Zaidi,[1] Rahul Bhandari,[2] and
Md Ehsan Asgar[3]

[1]Department of Computer Science and Engineering,
Panipat Institute of Engineering and Technology, Samalkha,
Haryana, India
[2]Schaefer School of Engineering and Science, Stevens Institute
of Technology, Hoboken, New Jersey, United States
[3]Department of Mechanical Engineering, HMR Institute of
Technology and Management, New Delhi, India

## 4.1 INTRODUCTION

The Internet of Things (IoT) is rapidly evolving, with numerous devices connected to the Internet every day. However, the rapid growth of IoT has led to security and privacy challenges, as well as interoperability and data exchange issues. Blockchain, a decentralized ledger technology, has the potential to revolutionize IoT by addressing these challenges and improving the reliability and security of IoT applications. This study comprehensively reviews the state-of-the-art efforts in using blockchain to revolutionize IoT. The introduction of the chapter highlights the growing importance of the IoT in connecting digital and physical objects to enable collaboration between computing systems, users, and objects. The IoT can potentially deliver significant convenience and economic benefits, such as optimizing supply chains, enhancing manufacturing processes, and improving healthcare delivery. However, the increasing number of connected devices concerns data security and privacy. As more devices are added to the network, ensuring that the communication between them is secure and that data privacy is maintained becomes more challenging. In this context, blockchain technology presents a promising solution to these challenges. Blockchain is a distributed ledger technology that maintains an immutable record of network transactions. It offers a secure, decentralized, and tamper-proof data management approach, making it suitable for addressing the security and privacy concerns of the IoT. This comprehensive review covers the most significant blockchain-based Internet of Things (B-IoT) applications, their architecture, and security considerations. Additionally, the challenges and future directions of B-IoT, which will guide researchers toward deploying the next generation of B-IoT applications, have been

DOI: 10.1201/9781003407096-4

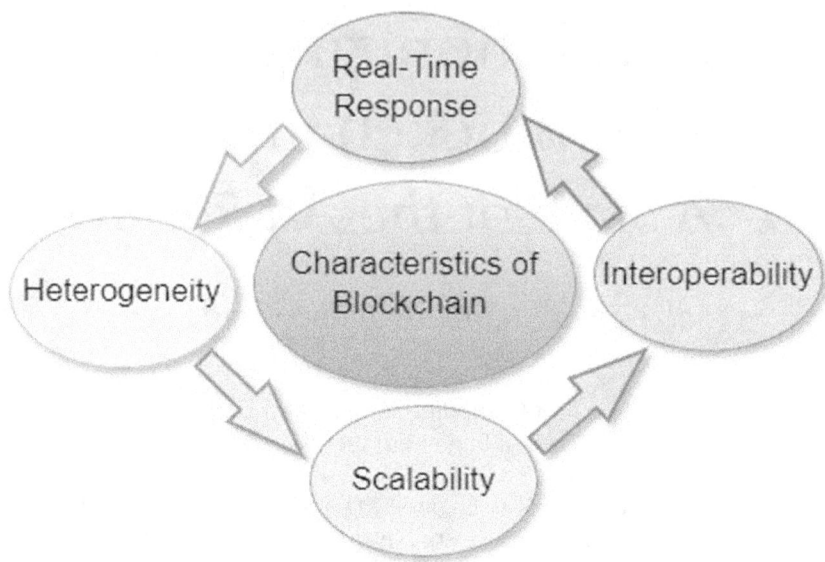

**FIGURE 4.1**   Characteristics of blockchain.

characterized and an overview of the chapter's objectives is provided and discussed in Figure 4.1. Overall, the introduction sets the stage for the chapter, highlighting the importance of the IoT and the need for secure and reliable communication between devices.

## 4.2   CHARACTERISTICS AND REQUIREMENTS

### 4.2.1   CHARACTERISTICS OF IoT

#### 4.2.1.1   Heterogeneity

IoT networks comprise a diverse range of devices, sensors, and networks often developed by manufacturers and operate on different protocols. This heterogeneity presents challenges in terms of interoperability and standardization. Heterogeneity is a key characteristic of IoT that presents significant challenges for achieving interoperability and standardization.

IoT systems are composed of devices and sensors with different technologies and standards that may have different communication protocols, data formats, and operating systems (Al-Fuqaha et al., 2015). This heterogeneity can make it difficult for devices to communicate with each other and with different networks and platforms, hindering the ability of IoT systems to realize their full potential. Research has proposed various approaches to manage the heterogeneity in IoT systems. For example, middleware has been identified as a potential solution to bridge the gap between devices and networks. Middleware acts as a layer between IoT devices and applications, providing a standard communication and data exchange interface. This approach can help address the interoperability challenge and facilitate the integration of heterogeneous devices and networks in IoT systems (Zanella et al., 2014). However, implementing middleware can introduce additional complexity, and selecting the appropriate middleware solution for a specific IoT application is crucial. There is no one-size-fits-all solution for middleware, and the middleware selection should be based on factors such as the required level of abstraction, scalability, security, and cost.

### 4.2.1.2  Scalability

IoT networks can accommodate many devices and services, making it possible to support complex systems. However, this also presents challenges in managing and securing such large and complex networks. Scalability is also an essential characteristic of IoT, and it refers to the ability of IoT systems to accommodate a growing number of devices and data without compromising performance (Li et al., 2015). As IoT systems become more widespread and connected, the amount of data generated by devices is expected to increase dramatically, which can present scalability challenges. Several solutions have been proposed to address scalability in IoT systems. One approach is to use distributed architectures that can distribute data processing and storage across multiple devices and nodes. This can help reduce the load on separate devices and enable more efficient data management (Gubbi et al., 2013). Another approach is to use cloud computing and edge computing technologies to offload some of IoT systems' processing and storage requirements. In this approach, IoT devices can send data to the cloud or edge computing infrastructure for processing and analysis, reducing the load on individual devices and enabling more efficient data management (Yi et al., 2015). However, as IoT systems continue to grow and generate more data, there is a need for continued research and development of new solutions to address scalability challenges. The IoT poses significant scalability challenges concerning the number of devices, the volume of data generated, the geographic distribution, and the diversity of applications and services that are envisioned (Karagiannis et al., 2016).

### 4.2.1.3  Interoperability

IoT systems should be designed to interoperate seamlessly with other systems, enabling cross-platform communication and collaboration. Standardization is crucial for achieving interoperability. Interoperability is another important characteristic of IoT, and it specifies the ability of different devices and systems to communicate and interchange data with one another (Zanella et al., 2014). In IoT, interoperability is essential for enabling devices from different manufacturers with different hardware and software specifications to work together seamlessly. The lack of interoperability in IoT can present significant challenges for developing and deploying IoT systems. Several interoperability standards have been proposed to address these challenges, including the ISO/IEC 30141 standard for IoT interoperability, which provides a framework for enabling interoperability across different IoT systems and devices (Yi et al., 2015). In addition, several initiatives and alliances have been formed to promote IoT interoperability, such as the Open Connectivity Foundation, which aims to develop and promote standards for IoT interoperability, and the Thread Group, which is focused on developing a standard for IP-based IoT networks (Khan et al., 2020). Despite these efforts, IoT interoperability challenges remain, including data formats, protocols, and security issues. Interoperability challenges are considered as one of the most significant issues that need to be addressed to figure out the potential of IoT.

#### 4.2.1.4   Real-Time Response

IoT systems must be able to respond in real-time, enabling timely decision-making and actions based on the data collected. This real-time response is essential for many applications of IoT, such as remote monitoring of patients. Real-time response is another important requirement of IoT, and it refers to the ability of IoT systems to process and analyze data in real-time, providing immediate feedback and response to changing conditions (Al-Fuqaha et al., 2015). This is critical for applications such as healthcare, transportation, and industrial control systems, where delays in data processing can have serious consequences. Achieving real-time response in IoT systems can be challenging, as it requires the efficient processing and analysis of large volumes of data in real-time. Several techniques and technologies have been suggested to address these obstacles, such as edge computing, which implies data processing at the edge of the network, nearer to the devices effectuating the data (Sun et al., 2020). In addition, real-time response in IoT can be facilitated by using machine learning and artificial intelligence algorithms, which can help identify patterns and anomalies in real-time data and trigger automated responses. However, there are still several challenges to achieving real-time response in IoT systems, such as data privacy and security issues and the need for more efficient and scalable data processing technologies (Chen et al., 2020).

## 4.3   REQUIREMENTS OF IOT

### 4.3.1   Security

The sensitive nature of the data that is exchanged between IoT devices makes security a top priority. Appropriate security measures must be in place to ensure the privacy and integrity of the data. Security is a crucial demand for IoT systems, as they involve exchanging sensitive information and controlling physical devices, which can have serious consequences if compromised (Li et al., 2015). IoT systems are particularly vulnerable to security threats due to their distributed nature, heterogeneity, and the large number of devices involved. A number of security protocols and mechanisms, including access control, encryption, and authentication, have been suggested to address these issues. Additionally, it has been suggested that blockchain technology could be used as a potential fix to increase the security of IoT devices (Wang et al., 2020). Blockchain is a distributed ledger technology that provides a tamper-proof and immutable record of transactions and has been proposed as a potential solution for securing IoT systems. By using blockchain, IoT systems can certify the integrity and confidentiality of data and develop secure authentication and access control mechanisms. However, many issues are still associated with using blockchain for securing IoT systems, such as scalability, energy efficiency, and interoperability. In addition, blockchain can introduce new security threats, such as 51% attacks and smart contract vulnerabilities (Dorri et al., 2017).

### 4.3.2 RELIABILITY

IoT networks must be designed to be highly reliable, with minimal downtime and low latency, to enable uninterrupted data exchange (Zanella et al., 2014). Reliability is another important requirement for IoT systems, as they often involve critical applications and services requiring high availability and uptime. Reliability signifies the ability of a system to operate consistently and predictably under various conditions and to deliver the expected performance levels. Several mechanisms and protocols have been proposed to ensure the reliability of IoT systems, such as redundancy, fault tolerance, and load balancing. These mechanisms aim to ensure the system can handle failures and adapt to changing conditions while maintaining the expected performance levels. By providing a tamper-proof and immutable record of transactions, blockchain technology can also increase the dependability of IoT systems. This can ensure that data is not lost or corrupted and provide an audit trail for troubleshooting and debugging. In addition, the distributed nature of blockchain can provide a high degree of fault tolerance and resilience, which can enhance the reliability of IoT systems. However, the use of blockchain can also introduce new challenges for the reliability of IoT systems, such as the need for synchronization and consensus among the nodes in the network. In addition, the computational overhead of blockchain can impact the performance and reliability of the system.

### 4.3.3 ENERGY EFFICIENCY

Many IoT devices are battery-powered and operate in low-power environments, so energy efficiency is essential. Energy efficiency is another important requirement for IoT systems, as many are battery-powered and operate in resource-constrained environments. IoT devices that consume a lot of energy can quickly drain their batteries, resulting in reduced performance and downtime. Several approaches have been proposed to ensure energy efficiency in IoT systems, such as power management, duty cycling, and sleep modes. These approaches aim to minimize the energy consumption of IoT devices while maintaining the expected performance levels (Zhang et al., 2018). Blockchain technology can also contribute to the energy efficiency of IoT systems by reducing the need for intermediaries and central authorities, which can consume a lot of energy. In addition, the distributed nature of blockchain can reduce the need for data duplication and can minimize the energy required for data transmission and storage. However, the use of blockchain can also introduce new challenges for the energy efficiency of IoT systems, such as the computational overhead required for consensus and validation. The energy consumed by the blockchain nodes and the communication overhead required for synchronization can also impact the system's energy efficiency (Chen et al., 2019).

### 4.3.4 CONNECTIVITY

IoT devices need to be connected to a network to enable communication and data exchange. The choice of connectivity protocol depends on the particular uses and requirements. Connectivity is an important requirement for IoT systems, as these systems rely on network connectivity to transmit data and interact with other devices (Li et al., 2015). However, IoT systems often operate in diverse and complex environments, challenging network connectivity. Several technologies have been proposed to address the connectivity challenges of IoT systems, such as Zigbee, Wi-Fi, Bluetooth, and cellular networks. Each of these technologies has its own strengths and weaknesses, and the choice of a technology depends on the particular requirements of the IoT system (Miorandi et al., 2012). Blockchain technology, in addition to conventional network technologies, can help IoT devices to be connected. By providing a decentralized and distributed network, blockchain can directly exchange data and transactions between devices without relying on centralized intermediaries. However, the use of blockchain in IoT systems can also introduce new challenges for connectivity, such as the need for high-speed data transmission and low latency. In addition, the limited processing power and memory of IoT devices can also impact the system's connectivity.

### 4.3.5 PRIVACY

IoT networks must ensure that information is gathered, stored, and used to protect individuals' privacy concerns. This requires the implementation of appropriate privacy policies and practices. Privacy is a critical concern in IoT systems, as they involve collecting and handling sensitive information and data. Blockchain can potentially address some of the privacy challenges in IoT by establishing a safe and transparent way for data exchange and storage. Using a decentralized and peer-to-peer architecture, blockchain can ensure that a single entity does not control sensitive data, reducing the risk of data breaches and unauthorized access. Several studies have explored the use of blockchain for privacy-enhancing applications in IoT systems. For example, a study proposed (Pham et al., 2021) a blockchain-based framework for secure and privacy-preserving data sharing in IoT systems. The authors suggested that blockchain can establish a safe and transparent way for data exchange while enabling privacy-preserving data sharing and access control. Another study (Singh et al., 2020) explored the use of blockchain for privacy-preserving data sharing in IoT systems. The authors suggested that blockchain can enable efficient and secure data sharing while ensuring data privacy and confidentiality. A review article (Li et al., 2020) discussed the potential benefits and challenges of using blockchain for privacy-enhancing applications in IoT systems. The authors suggested that blockchain can enable secure and transparent data sharing while promoting privacy and confidentiality.

## 4.4 EVOLUTION OF BLOCKCHAIN

Blockchain is a distributed ledger system that enables safe and transparent transaction processing by eliminating the requirements of other intermediaries. It was initially designed for use in cryptocurrency, but its potential applications have expanded to other fields, including IoT. The core working principles of blockchain are auditability, security, and decentralization.

B-IoT can transform current infrastructure by integrating decentralized architecture, improved security, rapid transactions, and data ownership control. B-IoT uses blockchain technology to give IoT systems a transparent, impermeable, and trust-based foundation. By guaranteeing data integrity through its immutable nature, providing secure identity management, easing interoperability across various devices, and implementing strong consensus procedures to prevent assaults and illegal access, blockchain eliminates IoT risks. Additionally, the decentralized structure of blockchain technology increases system resiliency and lowers reliance on centralized authorities. Overall, B-IoT and blockchain technology can address serious flaws in current IoT infrastructure and transform it to be more efficient, secure, and trust-worthy. Some critical contributions and state-of-the-art are discussed in Table 4.1.

**TABLE 4.1**
**Important Findings and Contribution**

| Author | Main Contribution | Key Findings |
|---|---|---|
| Gubbi et al. (2013) | Sketch up an IoT vision, emphasizing its structural components and potential future developments. | Discusses the architectural components of IoT, like devices and protocols, as well as upcoming research fields like standardization, energy efficiency, cloud integration, and big data analytics. |
| Al-Fuqaha et al. (2015) | IoT in-depth study, focusing on supporting technologies, protocols, and applications. Provide a thorough overview of the Internet of Things, including its enabling technologies, protocols, uses, and difficulties. | Insights into different IoT protocols, cross-domain applications, and interoperability, security, scalability, and privacy difficulties. Discussions of IoT technology, communication protocols, various applications, and difficulties with security, privacy, interoperability, scalability, and energy efficiency are among the key results. |
| Khanna and Kaur (2020) | Describe the IoT generally, focusing on its features, uses, and related problems. | Characteristics of IoT, the variety of applications it has across disciplines and its difficulties in terms of privacy, security, data management, and standardization. |
| Kumar et al. (2020) | It involves challenges to attaining interoperability in the Internet of Things (IoT) domain, solutions, and future directions. | Provides insights into future perspectives for establishing seamless integration and collaboration among heterogeneous IoT devices and systems. Outlines major challenges in IoT interoperability and explores alternative techniques and technology to address them. |
| Pham et al. (2021) | IoT systems using a blockchain-based framework for safe and private data sharing | The framework creates a safe environment for data transmission by ensuring data integrity, enhancing security and privacy, and facilitating controlled data sharing. |

## 4.5 BLOCKCHAIN-BASED IOT APPLICATIONS

Several B-IoT applications have been proposed and are shown in Figure 4.2.

### 4.5.1 SUPPLY CHAIN MANAGEMENT

Using blockchain, supply chain can be tracked from end-to-end, improving transparency and reducing fraud. Blockchain can provide end-to-end transparency and
traceability in supply chains, helping to ensure the authenticity and integrity of
products and reduce the risk of fraud or counterfeiting. The use of smart contracts in
conjunction with blockchain can also automate supply chain processes and improve
efficiency. For example, a study (Xu et al., 2020) proposed a blockchain-based
supply chain management system for the agricultural industry, which was designed
to enhance transparency, traceability, and accountability in the supply chain. The
system used IoT sensors to collect data on the quality and condition of agricultural
products at different stages of the supply chain, which was stored on a blockchain
platform. Smart contracts were used to automate supply chain processes and enforce
rules and regulations related to food safety and quality. Another study (Azzi et al.,
2021) proposed a blockchain-based supply chain management system for the automotive industry, which was designed to improve traceability and reduce the risk of
counterfeit parts. The system used radio-frequency identification tags to monitor the
handling of parts throughout the supply chain, and stored this data on a blockchain

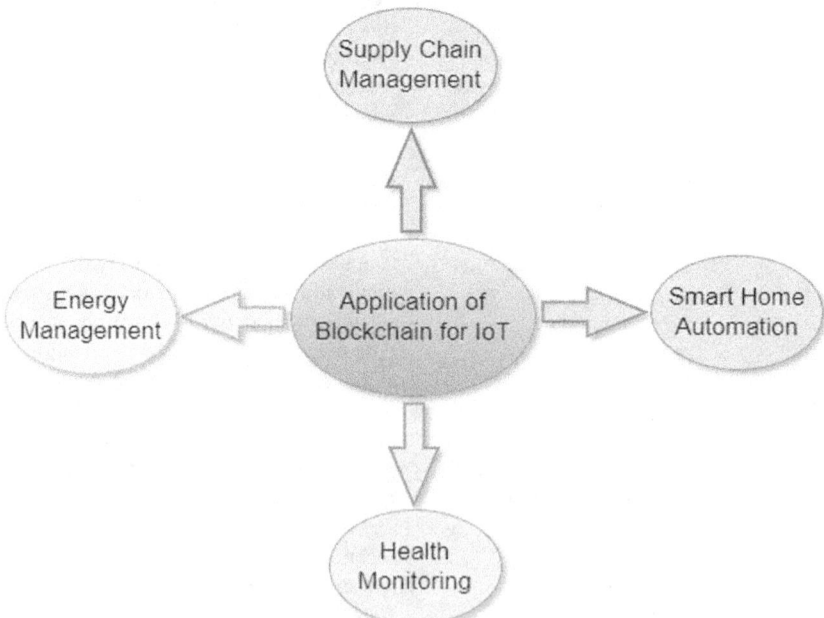

**FIGURE 4.2**   Application of blockchain for IoT.

platform. Smart contracts were used to automate the verification and certification of parts, and to ensure that only genuine parts were used in the production process.

### 4.5.2 Smart Home Automation

Blockchain can be utilized to develop a secure, decentralized network of smart home devices, improving privacy and security. Smart home automation is a popular application of IoT that can be revolutionized with blockchain technology. Smart homes consist of various connected devices that can be controlled through a centralized system. Blockchain can bring security and trust in smart homes by enabling secure communication, data privacy, and authentication. For example, a study (Ouaddah et al., 2018) proposed a blockchain-based framework for secure and decentralized home automation systems. The framework uses smart contracts to automate the communication and coordination between devices, while maintaining data privacy and secure authentication. Similarly, a study (Hao et al., 2018) proposed a blockchain-based smart home system that utilizes decentralized control and data sharing to enable secure and efficient smart home management. The system uses a distributed ledger to store data and smart contracts to automate various functions, including energy management and security. Blockchain-based smart home systems can provide high security and privacy while enabling efficient management of connected devices. Such systems can potentially revolutionize how we interact with our homes, bringing convenience and economic benefits to homeowners.

### 4.5.3 Health Monitoring

Blockchain can be used to store health data securely and can enable secure data sharing among healthcare providers and patients. Health monitoring is a key application of B-IoT that has attracted significant attention from researchers and healthcare practitioners due to its potential to enhance healthcare delivery and improve patient outcomes. In this application, real-time patient health monitoring and data transmission to healthcare practitioners are accomplished through the use of IoT devices and sensors. Blockchain technology can be used to protect patient privacy, secure the transmission and storage of health data, and verify the accuracy of the data. For instance, a recent study (Li et al., 2021) proposed a B-IoT-based health monitoring system that integrates IoT devices and blockchain technology to monitor and manage patients with chronic diseases. The system uses IoT devices to collect patient health data, including pulse rate, blood pressure, and sugar levels, and transmits it to a blockchain network. The blockchain stores and manages the information securely, and healthcare providers can access it in real-time to make informed decisions about patient care. Another study (Kumar et al., 2020) proposed a B-IoT-based healthcare system that leverages blockchain to certify the security, privacy, and accuracy of patient's health data. The technique uses IoT devices and sensors to collect patient's health data and transmit it to a blockchain network. The blockchain network stores the data securely and in a tamper resistant manner, will be better suited, and healthcare providers can access it to provide personalized patient care. Overall, B-IoT-based health monitoring systems can transform healthcare delivery by providing healthcare providers with real-time, safe, and accurate health data. To fully utilize this technology in healthcare, a number of issues like data privacy, interoperability, and legal compliance must be resolved.

### 4.5.4 ENERGY MANAGEMENT

Blockchain technology can track and manage energy usage across various devices and systems. This can help reduce energy waste, promote more efficient energy consumption, and create a more decentralized energy system. Energy management is another key area where B-IoT can bring about significant improvements. Integrating blockchain and IoT can make energy management systems more decentralized, transparent, and efficient. With B-IoT, devices in a smart grid can autonomously exchange energy based on pre-defined rules and agreements without requiring any intermediaries. This can lead to more efficient utilization of renewable energy sources and reduced energy waste. Blockchain technology also provides a tamper-proof and transparent way to monitor and record the origin of renewable energy sources, which is crucial for ensuring the integrity of renewable energy certificates (RECs). Using blockchain, RECs can be tracked and verified securely and transparently, which can enhance the credibility and marketability of renewable energy. Researchers explored the potential of blockchain for energy management in the context of B-IoT. For instance, the model suggested a blockchain-based energy management system for a microgrid, which uses smart contracts to facilitate energy transactions and ensure the integrity and security of the system (Li et al., 2019). Similarly, a blockchain-based energy trading platform enables peer-to-peer energy trading and improves the efficiency of energy markets (Raj et al., 2019).

## 4.6   ARCHITECTURE DESIGN AND SECURITY ASPECTS

The architecture of B-IoT applications depends on the specific use case, but a common design includes the use of smart contracts, which are the conditions of the contracts for both buyer and seller being directly encoded into lines of code and act as a self-executing agreement. Smart contracts enable safe, decentralized transactions by automating processes. Security aspects of B-IoT applications include protecting against attacks such as Distributed Denial of Service, 51% attacks, and Sybil attacks.

The architecture design and security aspects of B-IoT applications are critical to ensure the system's proper functioning and security. In B-IoT systems, the blockchain network typically consists of multiple interconnected nodes that maintain a shared ledger of transactions. The architecture design of a B-IoT system can vary depending on the specific use case, but it typically involves the integration of IoT devices and blockchain nodes to create a decentralized network. B-IoT systems' security is a significant concern since they handle sensitive data and control physical assets. The use of blockchain technology can improve system security by providing features such as immutability, auditability, and decentralization. However, B-IoT systems face several unique security challenges, such as attacks on IoT devices and vulnerabilities in the blockchain network. Researchers have proposed various security mechanisms to address these security challenges, such as authentication, access control, and encryption. For instance, the use of public-key cryptography can ensure secure communication between IoT devices and blockchain nodes, while access control mechanisms can restrict unauthorized access to sensitive data. Moreover, several consensus algorithms, such as Proof-of-Work, Proof-of-Stake, and Practical Byzantine Fault Tolerance, have been proposed to ensure the security and reliability of the blockchain network. These algorithms enable the nodes to reach a consensus on the validity of the transactions and prevent malicious actors from tampering with the ledger. Several research papers have focused on the architecture design and security aspects of B-IoT applications. Some model proposes a secure B-IoT system architecture that utilizes blockchain technology and lightweight cryptography. The proposed architecture addresses the security challenges of B-IoT systems, such as data privacy, authentication, and access control, by incorporating several security mechanisms (Aljawarneh et al., 2020). When blockchain technology and the Internet of Things (B-IoT) are combined, both potential problems are presented by architectural design and security issues. Scalability, interoperability, and decentralization are important considerations in B-IoT system architecture. Addressing any security flaws brought on by blockchain technology and establishing strong authentication and access control procedures are security problems. A crucial strategy is required to create safe, scalable, and interoperable B-IoT systems that maximize their promise while minimizing their hazards (Dagher et al., 2018; Kosba et al., 2016).

## 4.7 RECENT ADVANCES AND FUTURE DIRECTIONS

Recent advances in B-IoT research include the development of new consensus algorithms, such as Proof-of-Authority and Proof-of-Reputation, which can improve the efficiency and scalability of blockchain-based systems. Future directions include addressing the challenges of interoperability and scalability and developing new use cases for B-IoT applications. Since the adoption of blockchain technology for IoT is still in its infancy, issues still need to be resolved. Some recent advances have been made in this field to address these challenges. For example, researchers have suggested the use of lightweight consensus algorithms to reduce the energy consumption of B-IoT systems (Shi et al., 2021; Reshi IA et al.,2022). Similarly, some researchers have developed hybrid consensus algorithms to increase the scalability and security of B-IoT systems (Chen et al., 2020). As for future directions, there are still several areas where research is needed. One such area is the development of efficient consensus algorithms that can handle large volume of data generated by IoT devices. Another area is developing a lightweight encryption method that can be utilized to secure the data transmitted between IoT devices and blockchain networks (Shrestha et al., 2021).

Additionally, there is a need to develop privacy-preserving techniques that can protect the sensitive data generated by IoT devices. The creation of permissioned blockchains for better scalability and privacy, integration with AI for wise decision-making, and the investigation of novel consensus methods are only a few recent developments in coupling B-IoT. Future directions include edge computing, interoperability frameworks, secure hardware solutions, lightweight cryptographic systems, and tackling social and ethical issues. The full potential of the B-IoT in diverse fields will be unlocked via continued study and collaboration.

## 4.8  CONCLUSION

This chapter presents an extensive overview of how blockchain can be used to develop B-IoT applications. The study described the basic attribute and necessity of IoT and discussed the evolution of blockchain technology. Various researchers provide numerous B-IoT applications, such as supply chain management, smart home automation, health monitoring, and energy management, and discuss their architecture design and security aspects. Furthermore, the study highlighted the recent advances made toward solving the centralized IoT challenges and discussed the future directions that need to be considered to deploy the next generation of B-IoT applications. In conclusion, the study shows that blockchain technology can revolutionize the IoT industry by providing a secure, reliable, and decentralized network for various B-IoT applications. The potential of blockchain technology in addressing the security, privacy, scalability, interoperability, and reliability issues of IoT has also been characterized. However, some challenges still need to be addressed, such as the energy efficiency and connectivity issues. Therefore, further research is required to explore new solutions and strategies to overcome these challenges and realize the full potential of B-IoT applications.

## REFERENCES

Al-Fuqaha, A., Guizani, M., Mohammadi, M., Aledhari, M., & Ayyash, M. 2015. Internet of things: A Survey on Enabling Technologies, Protocols, and Applications. *IEEE Communications Surveys & Tutorials*, 17(4), 2347–2376.
Aljawarneh, S., & Hindy, H. 2020. Design of a Secure Blockchain-Based Internet of Things Architecture. *IEEE Internet of Things Journal*, 7(8), 7116–7124. doi: 10.1109/JIOT.2020.2991007
Azzi, R., Hasan, M. A., & Al-Qutayri, M. 2021. Blockchain-Based Supply Chain Management for Automotive Industry. *Future Generation Computer Systems*, 117, 257–268.
Chen, J., Jiang, J., Li, Z., Zhang, L., Chen, X., & Xu, Z. 2020. A Hybrid Consensus Algorithm for Blockchain-Based IoT Systems. *IEEE Access*, 8, 238321–238332.
Chen, M., Mao, S., & Liu, Y. 2020. Big Data for Mobile Health: A Review of Recent Advances. *IEEE Access*, 8, 182526–182544.
Chen, H., Xu, Y., Liu, L., & Hu, F. 2019. Energy-Efficient Blockchain for Internet of Things. *Future Generation Computer Systems*, 93, 797–808.
Dagher, G. G., Mohler, J., Milojkovic, M., & Marella P. B. 2018. Ancile: Privacy-Preserving Framework for Access Control and Interoperability of Electronic Health Records Using Blockchain Technology. *Sustainable Cities and Society*, 39, 283–297.
Dorri, A., Kanhere, S. S., Jurdak, R., & Gauravaram, P. 2017. Blockchain for IoT Security and Privacy: The Case Study of a Smart Home. In *2017 IEEE international conference on pervasive computing and communications workshops (PerCom workshops)* (pp. 618–623). IEEE.
Gubbi, J., Buyya, R., Marusic, S., & Palaniswami, M. 2013. Internet of Things (IoT): A Vision, Architectural Elements, and Future Directions. *Future Generation Computer Systems*, 29(7), 1645–1660. doi: 10.1016/j.future.2013.01.010

Hao, H., Wu, Z., Wu, J., & Hu, X. 2018. A Blockchain-Based Smart Home Management System. *Future Generation Computer Systems*, 86, 648–655.

Karagiannis, G., Besis, S., Papavassiliou, S., & Kouvelas, A. 2016. A Survey on Internet of Things: Architecture, Enabling Technologies, Security and Privacy, and Applications. *IEEE Communications Surveys & Tutorials*, 18(1), 161–190. doi: 10.1109/COMST.2015.2477041

Khan, Z., Anjum, A., Soomro, K., & Baik, S. W. 2020. Internet of Things (IoT) Interoperability: Challenges, Approaches, and Future Directions. *Journal of Network and Computer Applications*, 160, 102632. doi: 10.1016/j.jnca.2020.102632

Khanna, A., & Kaur, S. 2020. Internet of things (IoT), applications and challenges: a comprehensive review. Wireless Personal Communications, 114, 1687–1762.

Kosba, A., Miller, A., Shi, E., Wen, Z., & Papamanthou, C. 2016. Hawk: The Blockchain Model of Cryptography and Privacy-Preserving Smart Contracts. In *Proceedings of the 2016 IEEE Symposium on Security and Privacy (S&P)* (pp. 839–858). IEEE.

Kumar, P., Goyal, D., & Singh, R. 2020. Blockchain-Based Healthcare System Using IoT. In *2020 11th International Conference on Computing, Communication and Networking Technologies (ICCCNT)* (pp. 1–6). IEEE.

Li, X., Li, L., & Li, H. 2019. Blockchain-Based Energy Management System in Microgrid. *Future Generation Computer Systems*, 97, 670–677.

Li, J., Lu, R., Liang, X., & Shen, X. 2020. Blockchain for Privacy-Preserving and Secure Internet of Things: A Survey. *IEEE Communications Surveys & Tutorials*, 22(2), 1252–1284.

Li, S., Xu, L. D., & Zhao, S. 2015. The Internet of things: A Survey. *Information Systems Frontiers,* 17(2), 243–259.

Li, X., Zhang, H., Liu, J., Wei, X., & Zhao, Y. 2021. Blockchain-Enabled IoT for Chronic Disease Management. *Journal of Medical Systems*, 45(2), 1–8.

Miorandi, D., Sicari, S., & Pellegrini, F. 2012. Internet of Things: Vision, Applications and Research Challenges. *Ad Hoc Networks*, 10(7), 1497–1516.

Ouaddah, A., Abou Elkalam, A., & Ait Ouahman, A. 2018. Towards Blockchain-Based Secure and Decentralized Home Automation Systems. *Journal of Ambient Intelligence and Humanized Computing*, 9(3), 551–567.

Pham, D. D., Tran, A. B., Nguyen, T. H., Nguyen, N. T., & Dinh, T. Q. 2021. A Blockchain-Based Framework for Secure and Privacy-Preserving Data Sharing in IoT Systems. *Journal of Ambient Intelligence and Humanized Computing*, 12(7), 7233–7250.

Raj, R., Singh, S., Chatterjee, A., & Garg, S. 2019. Blockchain-Based Decentralized Energy Trading Platform for Smart Grids. *IEEE Transactions on Industrial Informatics*, 15(8), 4697–4705.

Reshi, I. A., & Sholla, S. (2022). Challenges for Security in IoT, Emerging Solutions, and Research Directions. *International Journal of Computing and Digital Systems, 12*(1), 1231–1241.

Shi, J., Zhang, Y., Jia, M., & Zhu, Q. 2021. Lightweight Consensus for Blockchain-Based IoT Systems. *IEEE Internet of Things Journal*, 8(1), 406–416.

Shrestha, P., Rijal, S., Pandey, S., & Kim, Y. 2021. Securing Blockchain-Based IoT: A Review of the State-of-the-Art. *Sensors*, 21(10), 3447.

Singh, T., Grewal, H. S., & Jha, S. K. 2020. Blockchain for Privacy Preserving Data Sharing in IoT Systems: A Survey. *Journal of Network and Computer Applications*, 164, 102797.

Sun, Y., Chen, Y., Li, S., & Zhao, X. 2020. A Survey of Edge Computing In IoT. *Journal of Industrial Information Integration*, 17, 100128. doi: 10.1016/j.jii.2019.100128

Wang, Y., Zhang, Y., Zhang, J., Zhu, L., & Zhang, Y. 2020. Blockchain-Based Security and Privacy in the Internet of Things: A Survey. *IEEE Internet of Things Journal*, 8(2), 1045–1060.

Xu, L. D., Chen, X., Li, S., & Wang, X. 2020. Blockchain-Based Supply Chain Management for Agriculture. *Computers & Electronics in Agriculture,* 176, 105572.

Yi, S., Li, C., & Li, Q. 2015. A Survey of Fog Computing: Concepts, Applications and Issues. *Proceedings of the 2015 Workshop on Mobile Big Data (Mobidata),* 37–42. (2018). ISO/IEC 30141:2018 Internet of Things (IoT) – Reference Architecture. doi: 10.1145/ 2808797.280880ISO/IEC, Retrieved from www.iso.org/standard/75875.html

Zanella, A., Bui, N., Castellani, A., Vangelista, L., & Zorzi, M. 2014. Internet of Things for Smart Cities. *IEEE Internet of Things Journal*, 1(1), 22–32. doi: 10.1109/JIOT.2014.2306328

Zhang, J., Li, Y., & Li, X. 2018. Energy-Efficient IoT: Architecture and Key Techniques. *IEEE Communications Magazine*, 56(12), 98–104.

# 5 A Framework for Smart and Resilient Supply Chains Based on Blockchain and the Internet of Things

*Hamed Nozari,[1] Javid Ghahremani-Nahr,[2] Maryam Rahmaty,[3] and Mahmonir Bayanati[4]*

[1]Department of Management, Azad University, Dubai Branch, UAE
[2]Faculty Member of Academic Center for Education, Culture and Research (ACECR), Tabriz, Iran
[3]Department of Management, Chalous Branch, Islamic Azad University, Chalous, Iran
[4]Faculty of Technology and Industrial Management, Health and Industry Research Center, West Tehran Branch, Islamic Azad University, Tehran, Iran

## 5.1 INTRODUCTION

Supply chain management includes processes in which a set of operations related to the supply of goods is planned. Supply chain management includes all movements of goods (logistics) and efforts to supply goods. Cloud computing and Internet of Things poses a significant impact on supply chain. The complexity of the supply chain causes challenges and defects in this field. The evolution of product supply processes can be considered a competitive advantage for technological companies (Ghahremani-Nahr et al., 2022).

Effective collaborative interactions with suppliers and adopting a win-win approach can effectively help solve new challenges in the industrial supply chain. On the other hand, it seems that blockchain technology, by emphasizing features such as unforgeability, traceability, immutability, decentralization, and transparency, can improve the current challenges of the supply chain, including quality, speed, and increasing the level of security, impenetrability, and resilience. The aim is to establish a system of cooperation and positive, cooperative interactions in the supply chain, improve various activities in order to plan, accelerate customer service, create operational and process synergy

DOI: 10.1201/9781003407096-5

between colleagues, respond faster to changing market demand, and ultimately produce diverse products optimally suited to customer needs. Information imported from the Internet of Things (IoT) technologies as a transformative technology in the context of blockchain technology is immutable, and other chain partners can track shipments, deliveries, progress, and transportation. Solutions based on the IoT and blockchain can assure end users about the authenticity and high quality of products and improve the level of satisfaction of all elements of the organization and consumers in different layers of the supply chain. Blockchain has the ability to reduce the amount of parallel and repetitive work and guarantees tracking. Apart from the obvious value of traceability, one can also benefit from the many benefits of reducing costs in the application of this technology (Chen et al., 2023).

Among the important parameters in collaborative supply chain interactions, we can mention joint creation of knowledge, knowledge sharing, and mutual trust between supply chain components. The combination of blockchain technology and the IoT in a proper way can create a suitable basis for collaborative interactions in the supply chain with the rapid circulation of correct, timely, and unmanipulated information (Hu et al., 2023). The combined technology of blockchain and the IoT, due to its remarkable features, can improve part of the current challenges of the supply chain, including improving collaborative interactions and its performance by improving quality, speed, delivery reliability, and resilience (Oudani et al., 2023).

In recent years, IoT and blockchain technologies have been used in many industries, such as food industries, pharmaceutical industries, automotive industries, and many other industries, and it has been proven that they can have many applications in operational processes and supply chain processes. With the increasing growth of the Internet and its applications in people's lives and some other advanced technologies in the last decade, we have seen important changes in supply chains. Efficiency and responsiveness are two important components of today's supply chains. Retailers also operate in a lean manner and maintain minimum inventory, and due to accurate demand forecasting, production planning and scheduling has become a complex task in the business environment. Therefore, in order to gain a competitive advantage, it is necessary to use up-to-date digital technology such as the IoT and blockchain in this field (Dedeoglu et al., 2023).

One of the areas that blockchain technology is expected to change is the supply chain and its related issues. Blockchain is considered an outstanding technology that is changing traditional business models and creating new opportunities in the entire supply chain. Therefore, this technology can increase the transparency, accountability, trust, security, and operational efficiency of the supply chain and reduce its costs. In addition, blockchain can be considered as a solution for tracking goods and information not only between manufacturers and suppliers but also throughout the entire supply chain. In the following, the dimensions and components of intelligent supply chains powered by the IoT and blockchain are examined more comprehensively.

## 5.2  INTERNET OF THINGS

The IoT is a relatively new approach that has grown increasingly in recent years and has created new strategies for wireless communication. The main idea of the IoT concept is the ubiquitous presence of objects around us and the possibility of communicating these objects through a wireless network. Radio-frequency identification (RFID) tags, all kinds of sensors, and mobile smartphones provide the possibility of integrated communication and all-around cooperation for all objects (Khan et al., 2023). The term "Internet of Things" was first introduced in 1999 by British scientist Kevin Ashton. Ashton proposed this concept in the form of a world where everything and every object has a digital identity, and computers control and manage them. In the IoT paradigm, many objects that surround us are networked in one or more forms. In today's world, various technologies based on wireless networks and sensors are being developed to overcome the challenges of the new era (with the presence of invisible communications). These technologies generate and store a large amount of data. To use these data, it is necessary that all of them are stored, processed, and presented in an integrated platform for decision-making.

Certainly, the IoT' main capability is its powerful impact on various aspects of daily life and the behavior of all actors in business processes. IoT applications can be seen in both business and personal life. In the field of daily and personal applications, convenience in lifestyle, health, advanced learning, entertainment, and reducing energy costs are just some of the possible application scenarios where the new paradigm will play a critical role soon. Similarly, from the perspective of business users, it can be seen to facilitate things such as automation and industrial production, logistics, business/process management, medical applications, health, security, and intelligent transportation of people and goods (Agrawal et al., 2023).

The IoT is like an ocean of big data that can help cities make accurate predictions, provide timely medical services, repair and maintain machinery, and help businesses generate insights from this data. With the progress of the IoT and its expanding applications, the volume of data generated by actuators and sensors has increased, and in this regard, it will provide many opportunities for businesses. Application of the IoT in production will lead to lower production costs and higher quality. Smart devices and sensors do not make mistakes and do their work accurately and efficiently. For those manufacturers who produce complex products such as aircraft components, ensuring that the components are connected seamlessly is very important. By using tracking technologies, in addition to avoiding additional costs, mistakes that are happening can be quickly prevented (Hassoun et al., 2023).

Using powerful data collection tools such as RFID systems, wireless smart networks, and various sensors and web networks based on cloud computing, the IoT identifies all the risks in the networks and reasonably deals with them. Cloud computing, using the powerful IoT infrastructure, integrates data and helps in the logical analysis of collected data. Since supply chains are increasingly complex in today's world, organizations consider these technologies as a tool to improve their technical and financial performance. These technologies are a powerful competitive advantage for organizations that want to stand out against others (Aliahmadi & Nozari, 2023). Using the IoT has many advantages for supply chain systems in organizations, such as:

1. Monitoring supply chain and transportation processes
2. Save time and money

3. Increasing employee productivity and monitoring and reducing errors
4. Integration and adaptation of business models
5. Make better business decisions
6. Generate more income
7. Ability to access information from anywhere and at any time
8. Improving communication between connected electronic devices
9. Transfer data packets through the network and save time and money
10. Automate tasks and help improve the quality of business services
11. Reducing the need for human intervention

The IoT encourages companies to rethink their business methods and provides them with the tools to improve their business strategy. In general, the IoT is used in manufacturing, transportation, and service organizations with an emphasis on the use of sensors in order to collect and store data. However, in recent years, these technologies have been used in the agricultural and food industries, infrastructure, homes, and smart cities. This ever-increasing growth has pushed some organizations towards transformative technologies.

The IoT, like artificial intelligence, has made our lives easier and brought us many benefits. The IoT can analyze the information as a skilled assistant and inform us of its results. The IoT gives us the feeling that we are dealing with a human being. This technology improves efficiency in the office, home, or urban environment.

## 5.3 BLOCKCHAIN

Blockchain technology is one of the latest and most attractive transformational technologies in business systems, creating powerful management changes. The purpose of blockchain is to create a transparent and decentralized structure. In the blockchain network, data is entered into databases in an interconnected structure called a block. The structural basis of each block is the previous block, and it is created on it. Each block has information that connects it to the previous block. These blocks form an information chain that is connected to each other in order of construction. The first block is called "Genesis Block."

Since blockchain technology works based on countless computers distributed worldwide, it can be considered a distributed ledger. This means that each block keeps a copy of the blockchain chain's data in its vicinity and cooperates with other blocks through it. The main part of any blockchain is its mining process. This extraction process is based on hash algorithms. The hash is the output of the math function. The input of this function can be any value, but its output is a unique value of fixed size. Hash functions are one-way, and their input cannot be obtained by having an output. The one-way hashing feature secures the blockchain network (Nozari & Nahr, 2022).

Blockchain-based systems secure themselves using powerful mechanisms, including advanced cryptographic techniques and mathematical decision-making behavioral models. One of the important features of blockchain-based programs is immutability and high security. This section discusses how to provide these two important features.

### 5.3.1 IMMUTABILITY

The consensus feature refers to the network nodes' ability to reach consensus, register transactions, and build blocks. On the other hand, immutability means preventing the copying of transactions already registered in the system. These two features together create security in the blockchain.

Consensus algorithms in the blockchain network assure us that the rules of the network are being implemented and that all actors in the network agree on the current state of the network. Immutability guarantees the integrity of data and maintains transaction records following the confirmation of each block.

### 5.3.2 BLOCKCHAIN ENCRYPTION AND SECURITY

Blockchain technology widely uses cryptography to secure the data in it. As mentioned before, hash functions are very important for cryptography. As stated in the previous section, hashing is a process in which a hash function takes an input of arbitrary size and returns a fixed-length hash output. In the hashing process, the output changes with the smallest change in the input. But if the input is fixed, the output will be the same no matter how often the function is executed.

In the blockchain space, the output functions (hash) are used as unique identifiers in the data block. The hash of each block is created using the previous block's hash. This is what creates the blockchain. Therefore, each block's hash depends on the previous block's hash and the data stored in the block. Hash identifiers play an important role in maintaining the security and immutability of the blockchain (Abdallah et al., 2023).

Blockchain has many uses. Blockchain can be used in almost any platform where there is a need to record and transfer data or messages. An efficient supply chain is the core of many successful companies whose purpose is to manage the distribution of goods and services from producer to consumer. Coordinating multiple stakeholders in a particular industry is very difficult using traditional methods.

One of the capabilities of blockchain technology is that it can create high transparency in businesses. A transparent, integrated supply chain ecosystem revolving around an immutable database is what many industries need to become more robust and reliable. Blockchain addresses precisely this need. Figure 5.1 shows the characteristics of the blockchain technology.

Blockchains do not have a central authority, which is a very efficient and scalable feature. Finally, blockchain can increase the productivity and transparency of supply chains and affect everything from warehousing to payments. Chain of command is necessary for many things; blockchain is the main link in the chain of command. Things that are necessary for reliability and integrity in the supply chain are provided by blockchain. Blockchain provides a consensus process to verify transactions. There are also no on-chain disputes regarding transactions because all parties on the chain have the same copy.

Blockchain creates integrity, and there will be no disputes about transactions in the chain because all the factors affecting the chain have the same copy of this digital ledger. In the blockchain, everyone can see an asset's ownership chain in the entire

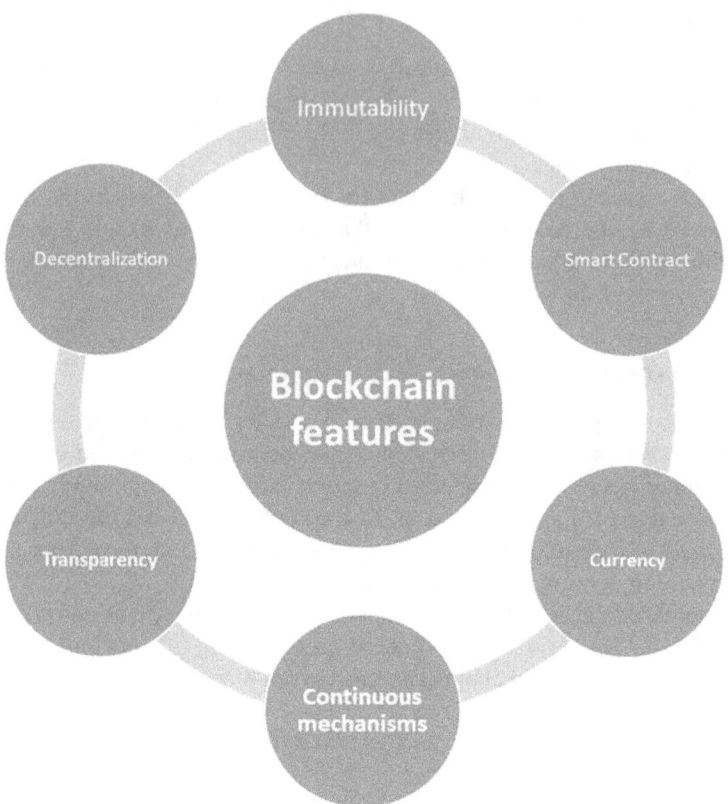

**FIGURE 5.1**   Features of blockchain technology.

chain. Records in the blockchain cannot be erased, and this issue is very important for the transparency of a supply chain.

## 5.4   SMART SUPPLY CHAINS

A smart supply chain includes using a variety of transformative and emerging technologies in the new era, including big data, IoT, and blockchain in the supply chain. These technologies allow supply chain-based organizations to optimize processes, reduce costs, shorten product delivery times, reduce negative environmental impacts, increase the sustainability of supply chain processes, and achieve high levels of automation (Aliahmadi et al., 2022).

A very important point about the connection between the IoT and the supply chain that gives rise to the intelligent supply chain is that such a supply chain is a self-improving and fully flexible system that can perform well in an unpredictable environment. To better understand what a smart supply chain is, you should know that such a system can process many things, including sales history, weather conditions, and the types of data it receives from its sensors, thus providing much better performance in logistics and supply chain (Nozari & Ghahremani-Nahr, 2022).

Automation of processes in the smart supply chain increases the speed of processes to a great extent. With the increasing growth of the IoT and blockchain technology and the formation of smart cities and businesses, the need to implement a smart supply chain in future businesses is strongly felt. In a smart business, all steps should be done intelligently, technologically advanced, and automatically. Therefore, as an advanced technology, the IoT, and blockchain play a very valuable role in the supply chain. With the development of the IoT and blockchain technology, supply chains are also becoming more modern and capable. Of course, various challenges in smart supply chains show that chain services still need continuous evolution and improvement. Anticipating needs in advance improves services (both to the organization's elements and consumers as end users). Information technology, blockchain, the IoT, and data processing are among the topics that make supply chain processes, from product supply to distribution and sales, improve and grow. Supply chain management is a multi-stage process in which different groups often operate, and each of them can somehow benefit from the role of the IoT in supply chain management. Among the most important of these advantages, the following can be mentioned (Azizi et al., 2021):

1.  Raw material suppliers use IoT-based solutions to track their processes. They can obtain information from real-time data about crop conditions and animal health in the agricultural industry and many other things. Continuing to process this information helps them increase their efficiency, improve the quality of raw materials and even reduce energy consumption.
2.  Solutions based on the IoT and the supply chain can be used to monitor production operations and equipment status in real-time. Continuous monitoring of tools and equipment makes it possible to detect failures in the shortest possible time, optimize the way of using assets and tools, and increase the efficiency of the production sector. IoT can also help improve sustainability.
3.  The IoT can bring more transparency and accuracy to the supply chain because logistics operators can receive real-time data about the location and status of each organization's assets. Using this data, they can modify and optimize the entire delivery route if changes are needed. IoT-based solutions also help to manage the supply chain of cold products, and by detecting problems along the way, they can keep perishable materials safe to a large extent.
4.  The IoT can facilitate the tracking of organizational inventories and significantly increase the accuracy of warehousing operations. Solutions that use the IoT and the supply chain can help preserve perishable goods by monitoring storage conditions.
5.  The IoT provides an overview of how supply chains affect a business, which can be key, especially in more complex supply chains. Providing an overview in this area can align different parts of an organization with each other and allow them to work together to prevent problems from occurring.

6. A supply chain in which the IoT is also used can help provide better customer service. Managers can access the information they need through mobile applications and accurately predict the delivery time. This issue can solve the problems related to the speed of product delivery and meet the customer's expectations.

7. With the use of Global Positioning System (GPS) displays today, it is possible for organizations and companies to track their sensitive goods in real-time and access their location at any moment.

8. By automatically collecting data, IoT-based systems and supply chains eliminate human data collection errors and help improve customer demand forecasting. Supply chain managers can forecast demand based on both historical and real-time data. Another critical point is that supply chain data can be collected nonstop or regularly. Either way, it allows businesses to tap into data that might be difficult or even impossible to collect manually.

The introduction of information and communication technology and the flow of information in the concept of the supply chain helps to integrate this chain better and faster. The integrated and virtual supply chain connects business partners through information and communication technology and enables the exchange of information needed for decision-making, including sales, purchase, product movement, financial flow, and service provision along the chain. In this case, not only is information shared throughout the supply chain, but it may also facilitate a company's internal operations or better cooperation between business partners in the supply chain using high-speed networks and databases. The effective use of these technologies is a key factor in the success of companies. The opportunities for new technology are increasing daily due to the ease and cheapness of access and connection to the global network. Data sharing provides opportunities that can be used to achieve significant effectiveness in the supply chain as well as a significant increase in customer service and responsiveness (Jayaraman et al., 2019). Collaborative interactions based on the IoT and blockchain in the supply chain will reduce the duration of product delivery to the market, reduce distribution time, and improve the quality of the organization's performance. In addition, technologies reduce the cost of delivery and shipping of products by reducing transportation and warehousing costs. These technologies reduce administrative costs and inventory and improve operational processes by reducing resource wastage. Sharing the information obtained from the IoT in the interaction between chain members and joint cooperation will improve the skills and abilities of all the elements involved in the organization. This will increase the efficiency and effectiveness of the use of facilities and resources in the entire value chain of the organization. It can also be that this cooperation across the chain, enhanced by the emphasis on the capabilities of the IoT and blockchain, can overcome all existing barriers and enable the use of new opportunities by sharing skills and resources. In addition, all actors involved in supply chain processes learn from each other by emphasizing the capabilities of smart systems based on artificial intelligence. So companies learn a lot about themselves and gain flexibility and transparency through collaboration. Increasing income, reducing costs, and sharing risk are among the things that provide the basis for strengthening the financial base of chain partners (Balamurugan et al., 2021).

Blockchain has great potential to reduce costs in supply chain processes and increase the level of service. For example, the following modes show these capabilities:

- *Identification of resources*
  Identifying sources is one of the most important areas of blockchain technology in various business fields. By using this functionality, companies can assess the risk level of suppliers as well as their credibility before making a decision. Since all interactions between suppliers and purchasing organizations exist in the supply chain processes, therefore, by using this blockchain capability, all performance can be monitored. In this case, the best decision can be made.
- *Tracking and interception*
  Lack of information transparency and lack of coordination among supply chain processes and their constituent elements are always there. This lack of coordination can disrupt the tracking of physical and financial interactions. Active elements in the supply chain processes always need information such as security and technical signs. The integration of responding to these needs has many challenges. By using systems based on blockchain and the IoT, all information, physical and financial transfers can be recorded. And they are intercepted.
- *Financial transactions and payments*
  Blockchain technology can be used for the growth and excellence of financial transactions and all types of payments in the supply chain. This will be much more efficient and powerful when financial supply chain systems are activated.

Figure 5.2 shows the smart supply chain ecosystem. The ecosystem approach in the supply chain, first of all, means understanding the supply chain as an integrated whole with different and related elements and understanding the connections between them. In other words, the supply chain ecosystem is a network of entities in various industry and business sectors that work together to design, and implement supply chain solutions and processes.

In general, it can be seen that among the most important advantages of establishing cooperation in the supply chain based on the IoT and blockchain are the significant reduction of costs, time reduction, and risk reduction in the production of new products. Continuous growth of processes, increase in profitability, transfer of experiences to future related activities, increase of innovative capabilities of all involved partners, sharing of knowledge and experience, reduction of risks, and overall growth of resilience are other benefits of using these combined smart systems. Blockchain-based solutions can give consumers more confidence in original and high-quality products and, to a significant extent, create a greater desire to buy that brand (Song et al., 2021).

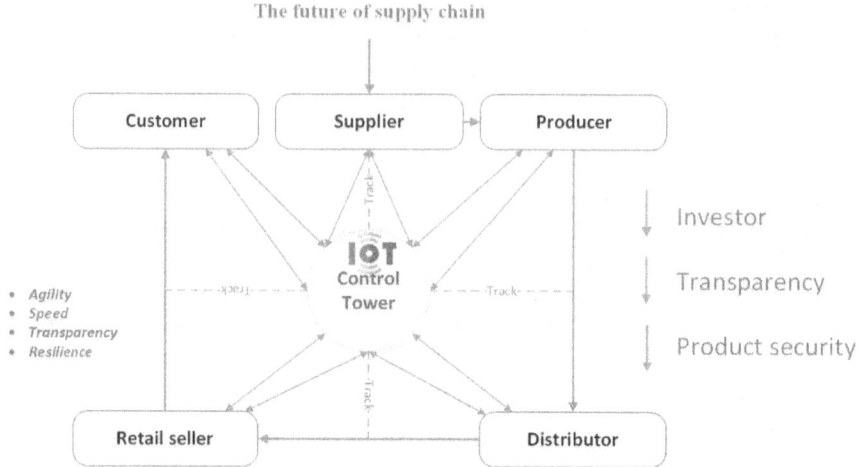

**FIGURE 5.2**   Smart supply chain ecosystem.

## 5.5   SMART AND RESILIENT SUPPLY CHAINS BASED ON BLOCKCHAIN AND IOT

Modern day supply chains around the world have become complex processes with many parties involved. For this reason, supply chain companies intend to improve their business operations using technologies such as the IoT and blockchain in order to monitor assets accurately. IoT sensors are used to collect information related to environmental conditions, verifying how long the cargo stays in a certain truck or a certain port, and whether there have been any violations or problems with the cargo or not. This information provides us with reassuring evidence that can be used in case of any invoice discrepancies. Companies can use this information to optimize their supply chain operations to increase their business productivity (Nozari & Aliahmadi, 2022).

The combined use of these two technologies, that is, IoT and blockchain technology, can bring high added value to businesses and supply chain processes. Some of these applications are presented here.

### 5.5.1   TRACK ALL ACTIVITIES THROUGHOUT THE SUPPLY CHAIN

Both organizations and end users can track entire product life cycle processes in the supply chain using IoT and blockchain technologies simultaneously. Blockchain is a detailed record of data in which all communications between devices equipped with the IoT are stored in its history. It is possible to access all information related to products quickly.

### 5.5.2   SMART CONTRACTS

Blockchain and the IoT can ensure cargo's financial and physical security even in long-distance transactions. This powerful integrated hybrid technology regulates the communication and interactions of all actors and actors logically involved in supply chain processes. In this way, the basic terms and conditions of the contract are rewritten in the form of encrypted computer codes, and financial transactions between all parties are carried out entirely anonymously, efficiently, and transparently without any conflicts.

Another advantage of the simultaneous use of blockchain and IoT technologies in business processes is the minimization of additional and tedious administrative procedures or, in other words, common bureaucracies. There are many intermediaries in the supply chain processes, from supply to distribution and sales. Since most organizations still use traditional business processes, the need for tedious paper-based procedures still exists in businesses. Blockchain and IoT technologies enable supply chain-reliant organizations to save money and time by eliminating tedious administrative procedures while ensuring better data flow protection across the ecosystem (Banerjee et al., 2019).

Because blockchain is a decentralized distributed ledger, no central authority or specific executive organization records data and transactions. Blockchain technology is designed based on a cryptographic algorithm so that we can use it to ensure the security of data and its immutability. Each block in the blockchain network has a hash of the previous block, which indicates that none of the blocks can be changed or replaced. Integrating blockchain with the IoT will help logistics companies deliver shipments more quickly and easily and optimize the operational costs of various applications. Figure 5.3 shows the smart supply chain framework based on the IoT and blockchain (Zhang et al., 2020).

Blockchain and IoT can improve supply chain management. Just imagine what benefits the transportation process can offer us by implementing these two technologies together.

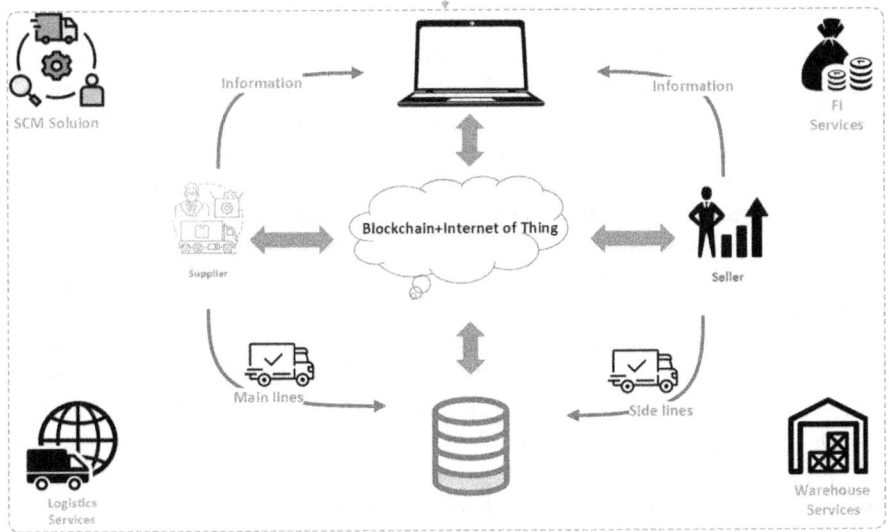

**FIGURE 5.3** The framework of the smart supply chain based on the B-IoT.

In some goods that need continuous and unchanged distribution, "keeping data and information in a safe place" will be very effective and practical. The secret of reaching these factors lies in blockchain's main and practical characteristics, which blockchain technology can play a revolutionary role in the supply chain. It is essential to mention that the use of the IoT and artificial intelligence technologies in the supply chain has been associated with high costs, but the blockchain and its infrastructure for companies active in this field are associated with cost reduction while being highly efficient. Blockchain efficiency makes global supply chains more efficient and resilient by allowing companies to complete and confirm transactions directly. Integrated payment solutions reduce the time between order and payment processing and ensure proper and timely movement of products.

## 5.6 CONCLUSION

The ever-increasing advancements in information technology capabilities have radically changed the face of the industry compared to the last decade. Acceptance and implementation of transformative technologies in the new era is one method that creates distinctive competitive ability in supply chain organizations and processes. In the past, the simplicity of supply chains was due to the locality of businesses, but with the development of marketing-related matters at global levels, the complexities of supply chains are increasing daily. As a result, there must be a tremendous change in the management of these chains.

IoT technology allows the surrounding objects to exchange information virtually, and by creating synergy, they cause significant growth in the quality of human life. On the other hand, organizations are looking for transparency, agility, and resilience in their supply chain to monitor and control the challenges and problems in the supply chain, which can be achieved using technologies such as blockchain and the IoT. This research presents a model of the Internet of user-oriented objects based on information through the interaction between the IoT and blockchain. This analytical framework creates the possibility of separate discussions regarding network, computing, storage, and visualization and, as a result, provides the opportunity for the growth of each independently but complementary to each other in a shared environment. Using the capabilities of IoT technology and blockchain, organizations' supply chain and logistics management problems, such as distribution network regulation, distribution strategy, information circulation, inventory management, and cash flow, are solved through a cloud-based vision. Cloud-based technologies facilitate collaboration with partners and customers by sharing information resources, leading to improved productivity and increased innovation.

The implementation of systems based on the simultaneous use of the IoT technology along with blockchain facilitates forecasting and planning processes and many sub-processes by using accurate, regular, and real-time information coverage. The supply chain based on these technologies allows organizations to spend their capital in production and operational processes instead of paying large costs in the software sector for higher security, which leads to more cash flow in business processes.

# REFERENCES

Abdallah, S., & Nizamuddin, N. (2023). Blockchain based solution for pharma supply chain industry. *Computers & Industrial Engineering*, 177: 108997.

Agrawal, P., & Narain, R. (2023). Analysis of enablers for the digitalization of supply chain using an interpretive structural modelling approach. *International Journal of Productivity and Performance Management*, 72(2), 410–439.

Aliahmadi, A., Nozari, H., & Ghahremani-Nahr, J. (2022). A framework for IoT and blockchain based on marketing systems with an emphasis on big data analysis. *International Journal of Innovation in Marketing Elements*, 2(1), 25–34.

Aliahmadi, A., & Nozari, H. (2023, January). Evaluation of security metrics in AIoT and blockchain-based supply chain by neutrosophic decision-making method. *Supply Chain Forum: An International Journal*, 24(1), 31–42.

Azizi, N., Malekzadeh, H., Akhavan, P., Haass, O., Saremi, S., & Mirjalili, S. (2021). IoT–Blockchain: Harnessing the power of internet of thing and blockchain for smart supply chain. *Sensors*, 21(18), 6048.

Balamurugan, S., Ayyasamy, A., & Joseph, K. S. (2021). IoT-Blockchain driven traceability techniques for improved safety measures in food supply chain. *International Journal of Information Technology*, 1–12.

Banerjee, A. (2019). Blockchain with IOT: Applications and use cases for a new paradigm of supply chain driving efficiency and cost. In *Advances in Computers* (Vol. 115, pp. 259–292). Elsevier.

Chen, X., He, C., Chen, Y., & Xie, Z. (2023). Internet of things (IoT)—blockchain-enabled pharmaceutical supply chain resilience in the post-pandemic era. *Frontiers of Engineering Management*, 10(1), 82–95.

Dedeoglu, V., Malik, S., Ramachandran, G., Pal, S., & Jurdak, R. (2023). Blockchain meets edge-AI for food supply chain traceability and provenance. *Comprehensive Analytical Chemistry* 101: 251–275.

Ghahremani-Nahr, J., Aliahmadi, A., & Nozari, H. (2022). An IoT-based sustainable supply chain framework and blockchain. *International Journal of Innovation in Engineering*, 2(1), 12–21.

Hassoun, A., Kamiloglu, S., Garcia-Garcia, G., Parra-López, C., Trollman, H., Jagtap, S., & Esatbeyoglu, T. (2023). Implementation of relevant fourth industrial revolution innovations across the supply chain of fruits and vegetables: A short update on traceability 4.0. *Food Chemistry*, 409, 135303.

Hu, H., Xu, J., Liu, M., & Lim, M. K. (2023). Vaccine supply chain management: An intelligent system utilizing blockchain, IoT and machine learning. *Journal of Business Research*, 156, 113480.

Jayaraman, R., Salah, K., & King, N. (2019). Improving opportunities in healthcare supply chain processes via the Internet of things and blockchain technology. *International Journal of Healthcare Information Systems and Informatics (IJHISI)*, 14(2), 49–65.

Khan, S., Haleem, A., Husain, Z., Samson, D., & Pathak, R. D. (2023). Barriers to blockchain technology adoption in supply chains: The case of India. *Operations Management Research*, 1–16.

Nozari, H., & Aliahmadi, A. (2022). Lean supply chain based on IoT and blockchain: Quantitative analysis of critical success factors (CSF). *Journal of Industrial and Systems Engineering*, 14(3), 149–167.

Nozari, H., & Ghahremani-Nahr, J. (2022). Assessing key performance indicators in blockchain-based supply chain financing: Case study of chain stores. *International Journal of Innovation in Engineering*, 2(3), 42–58.

Nozari, H., & Nahr, J. G. (2022). The impact of blockchain technology and the internet of things on the agile and sustainable supply chain. *International Journal of Innovation in Engineering*, 2(2), 33–41.

Oudani, M., Sebbar, A., Zkik, K., El Harraki, I., & Belhadi, A. (2023). Green blockchain based IoT for secured supply chain of hazardous materials. *Computers & Industrial Engineering*, *175*, 108814.

Song, Q., Chen, Y., Zhong, Y., Lan, K., Fong, S., & Tang, R. (2021). A supply-chain system framework based on internet of things using blockchain technology. *ACM Transactions on Internet Technology (TOIT)*, *21*(1), 1–24.

Zhang, H., & Sakurai, K. (2020). Blockchain for IoT-based digital supply chain: A survey. In *Advances in Internet, Data and Web Technologies: The 8th International Conference on Emerging Internet, Data and Web Technologies (EIDWT-2020)* (pp. 564–573). Springer International Publishing.

# 6 Pharma-Blocks
## *Blockchain-IoT Platform for Pharmaceutical Sector*

*Adil Mudasir Malla and Asif Ali Banka*

Department of Computer Science and Engineering,
Islamic University of Science & Technology, Awantipora,
Kashmir, J&K, India

## 6.1 INTRODUCTION: BLOCKCHAIN TECHNOLOGY

"Blockchain" refers to recording and validating financial transactions in a linked block network [1]. An information block is a set of transactions validated in the network and the two consecutive blocks are linked cryptographically. A blockchain is a distributed, immutable, append-only database across a network of computers that holds information blocks for use in safe, peer-to-peer financial transactions that use cryptography. The transactions depict state change, for instance, the transfer of coins from the sender to the beneficiary, where the account balance represents the transaction state. A miner node (computer) collects broadcasting of new transactions in the network and combines them with other transactions to create a block & broadcast it across the network. All valid transactions are verified before they get recorded in a blockchain. A block with the proper cryptographic hash can be considered valid and mined [2]. The *blockchain consensus* based on this collection of principles is the process by which all participants in a blockchain network agree on a single state of the blockchain as the network's verifiable reality. Since attempting to double-spend a transaction causes a conflict with the current state, the consensus mechanism prevents it by rejecting the transaction and never adding it to the chain. A *consensus protocol* is a set of rules for validating transactions, adding newly generated blocks, and choosing which fork or partition to use if the network splits. The circumstances and use cases will determine the necessity and requirement for a consensus.

A *smart contract* component of the blockchain is a piece of code with programmable application logic that runs itself, cannot be changed, and can be checked by both parties. It lets people make their application logic, which is then carried out by the blockchain, like a law. It means that third parties are no longer needed to help with transactions or verify who owns an asset. Intelligent contracts form the basis of trust at the application layer. The distributed ledger system (blockchain, umbrella term) allows users to share transaction records that have been checked and are up to date. All nodes share this information in the network. A coordinating entity is no longer needed to perform validation checks. Cryptocurrencies like Bitcoin are similar to fiat

DOI: 10.1201/9781003407096-6

currencies like the US dollar or the Euro in that they enable trade-in value. However, their decentralized governance relies on cryptographic protocols instead of a central agency (bank). Bitcoin [3] uses blockchain, an open peer-to-peer value transfer network that solved the double spending issue for the first time. However, blockchain technology's applications extend far beyond digital currency and are a powerful tool for developing trustworthy products and services.

## 6.2   OPPORTUNITIES AND CHALLENGES IN THE ADOPTION OF BLOCKCHAIN TECHNOLOGY

Practitioners and researchers alike have noticed that there have been a lot more studies and projects about blockchain in the past few years. Blockchain's distributed database makes it possible to keep track of transactions in a way that's both secure and easy to track back to their original owners and auditors. Furthermore, a distributed database can manage a growing set of documents called nodes [1]. The main advantage of blockchain technology is the immutability of contracts and transactions once they have been recorded in the distributed database [4]. The advantages and disadvantages of using blockchain technologies are shown in Figure 6.1.

### 6.2.1   WHY BLOCKCHAIN IN PHARMACEUTICAL INDUSTRIES?

Maintaining reliable and fast operations and a secure supply network is critical in the pharmaceutical industry. Many pharmaceutical industry stakeholders lose money due to theft and misplacement of goods due to a lack of reliable monitoring and tracing [5]. Officials in the pharmaceutical and healthcare industries have also discovered counterfeit and poorly made goods for sale. Many pharmaceutical companies have investigated using blockchain technology to centralize data, facilitate transaction tracking, and strengthen supply chain security in response to malpractice and inefficient supply networks. Figure 6.2 illustrates the role of blockchain in pharmaceutical industries. Figure depicts the blockchain solving different pharmaceutical challenges. Also the present scanerios of pharmaceutical businesses are presented.

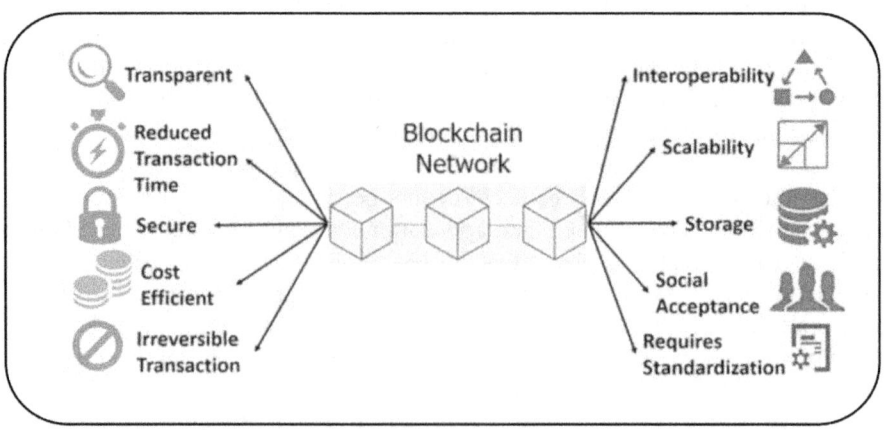

**FIGURE 6.1**   Opportunities vs. challenges in the adoption of blockchain.

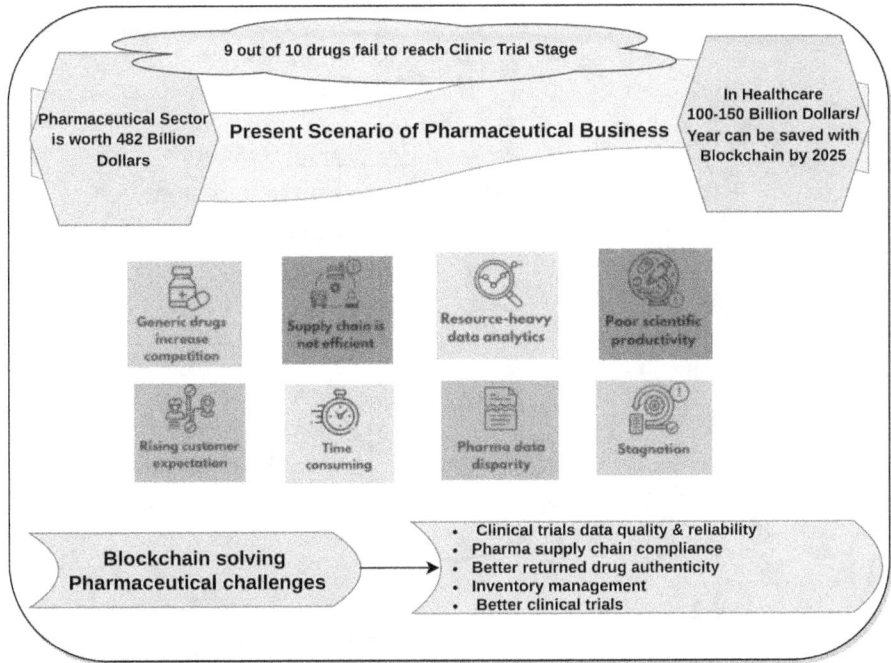

**FIGURE 6.2**   Blockchain in pharmaceutical industries.

## 6.3   BLOCKCHAIN TECHNOLOGY IN THE PHARMACEUTICAL COMPANIES

Companies in the pharmaceutical industry are continuously researching and developing new medicines to serve patients better. These drugs go through a multi-year, multi-stage procedure, beginning with discovery/pre-clinical research and ending with commercialization, to ensure patent protection, efficacy, safety, regulatory clearance, and validity, as shown in Figure 6.3. Such a long/protracted process is unsafe for drug recall because of the absence of privacy and security it entails [6]. Among the obstacles that the conventional pharmaceutical supply chain must overcome are more transparency, difficulty in tracking products, a lack of trust, and the shipment of expired products [7]. However, the pharmaceutical company must maintain accurate documents of the origins of all raw materials used in producing and distributing their products [5, 8, 9, 10]. Counterfeit drugs are inherently fraudulent and manufactured outside the approved pharmaceutical production system. Since counterfeit medications resemble the genuine product so closely, they are frequently challenging to detect. To the uninitiated eye, they appear identical to the genuine article and are unlikely to have any discernible adverse effects on the body. However, they do not always function as intended and may contain hazardous or toxic substances. In addition, in both therapeutic and preventive contexts, medicines are used. In recent years, the increase in counterfeit drug manufacturing has also impacted industries that manage

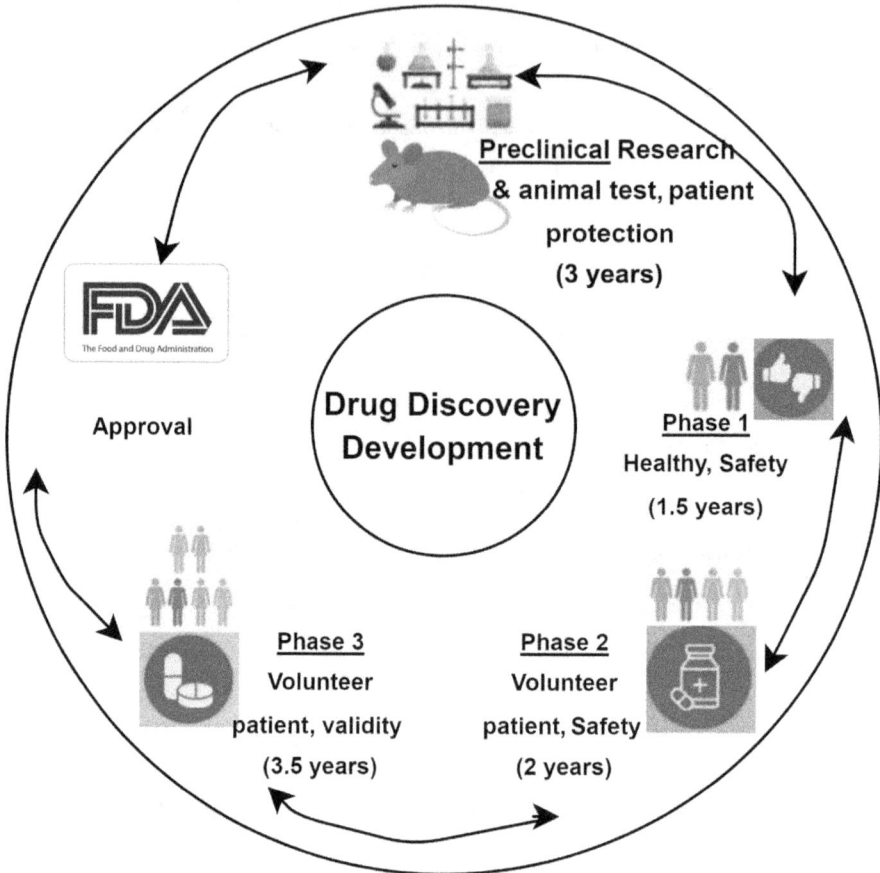

**FIGURE 6.3**   Drug discovery development in pharmaceutical companies.

confidential information, such as the military, healthcare, banking, and finance [11, 12, 13].

Security is a significant concern in the pharmaceutical industry; cryptographic technologies verify blocks of transaction records [7]. Standardization combats the safety risk of counterfeit medications [14]. This innovation verifies serial numbers throughout the supply chain. From the manufacturer to the pharmacy, blockchain chaincodes, data miners, and health information maintain quality [9]. Drug theft has decreased due to enhanced drug traceability [15].

Fernando et al. reviewed the literature and found that transparency, traceability, trust, tracking, and real-time analysis are needed for blockchain technology to work in the pharmaceutical industry. A review [15] examined how the pharmaceutical business could use blockchain technology. Researchers in Saudi Arabia found that healthcare employees' attitudes, economic inequality, and a lack of collaboration were the most significant challenges to implementing blockchain technology. Additional factors that they believe may help blockchain apps come into play are system robustness,

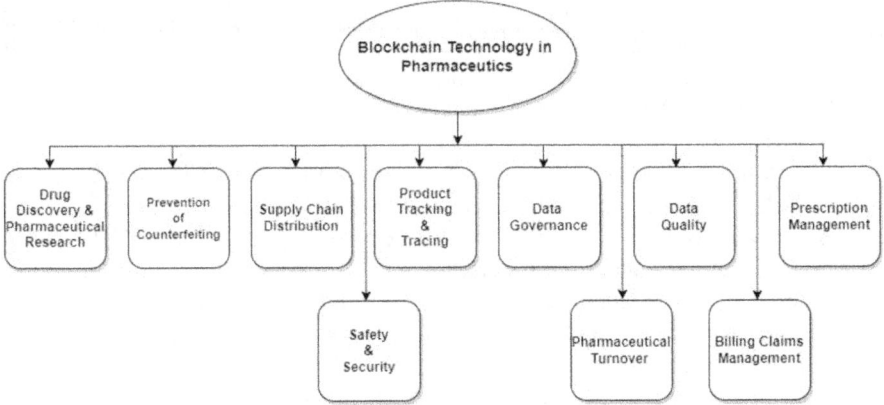

**FIGURE 6.4** Blockchain in pharmaceutical industries.

decentralization, data safety, interoperability, enhanced supply chain management, and government policies and laws. Traceability issues in the pharmaceutical supply chain were reviewed by Uddin et al. along with strategies for using blockchain technology in tracking and tracing to cut down on fake drugs. Notably, none of the above research went beyond a high-level summary of blockchain's potential uses [4,16].

The latest studies describe blockchain in the pharmaceutical sector investigated for the various subfields of drug development, as depicted in Figure 6.4.

### 6.3.1 DRUG DISCOVERY AND PHARMACEUTICAL RESEARCH

As depicted in the previous two sections, pharmaceutical research could lead to the development of innovative medicines for various diseases. The drug design and development, including its corresponding research, is a time-consuming process, so utmost care is taken once a drug is developed for the protection of its innovation along with its clinical research as described below.

#### 6.3.1.1 Intellectual Property (IP) Administration

IP is a critical aspect of the pharmaceutical sector, as it protects the innovation and research of pharmaceutical companies. However, managing IP can be complex and challenging, involving multiple stakeholders, patent applications, and regulatory compliance. Blockchain technology can offer a secure and transparent way to manage IP in the pharmaceutical sector, providing a tamper-proof and immutable record of all IP-related transactions.

One way blockchain is incorporated in IP management is by designing a digital ledger that records all IP-related transactions. This ledger can include patent applications, trademarks, copyrights, licensing agreements, and the parties' identities in each transaction. Authorized stakeholders, such as researchers, patent offices, and regulatory bodies, can access the ledger, providing a transparent and secure way to manage IP. Blockchain can also streamline the patent application process, reducing the time and cost of obtaining patents. For example, smart contracts automate the patent application process, ensuring all required information is submitted correctly and promptly. It helps to accelerate the process of obtaining patents while reducing the risk of errors and delays. Overall, blockchain can be incorporated into IP management of pharmaceutical drugs to yield benefits in the following ways:

*Secure storage:* Blockchain securely stores digital records of IP, such as patents and copyrights, which can help to prevent tampering and counterfeiting.

*Verification:* Blockchain verify the authenticity of digital records of IP, ensuring that they have not been altered or duplicated.

*Smart contracts:* Blockchain can create smart contracts that automatically enforce IP rights, such as royalties and licensing fees. In this way, it reduces the cost and complexity of IP management.

*Transparency:* Blockchain creates a transparent and auditable record of all IP-related transactions, including licensing agreements, royalties, and ownership transfers. So, it prevents disputes and ensures inventors and creators are appropriately credited and compensated.

*Decentralization:* Blockchain creates a decentralized platform for managing IP, which can help reduce intermediaries' influences and provide more direct and fair compensation to inventors and designers.

*Open innovation:* Blockchain can create an open platform for innovation where inventors and designers can collaborate and share their ideas while protecting their IP rights. It helps to accelerate innovation and reduce the cost of research and development.

Numerous blockchain-based IP administration solutions in the generic space aid in drug development innovations. The desire to use Labii's blockchain-supported electronic lab notebook solutions is an example. Bernstein provides blockchain-based management of digital trails with time-stamping to safeguard IP precedence, which is helpful in collaborative pharmaceutical research. A solution from iPlexus uses blockchain to make all published and unpublished data from drug development studies readily available. Blockchain solves the puzzle of keeping trust and protecting IP to allow for such an innovative initiative and framework.

### 6.3.1.2 Clinical Studies

Another potential use case for blockchain in IP management is clinical trial data management. Clinical trial data is a critical aspect of the pharmaceutical industry and is often subject to regulatory compliance and IP protection. Blockchain creates a secure and transparent ledger of all clinical trial data, ensuring that it is protected and easily accessible to authorized stakeholders.

The IEEE Standard Association held a session on Blockchain for Clinical Studies to speed up drug discovery, improve patient recruitment, and ensure data integrity. Scrybe, a blockchain initiative that provides a reliable and fast mechanism for speeding up clinical trials and research, was introduced at the forum [16]. It offers a clear and straightforward framework for inspectors to assess the trial's adherence to legal and ethical standards. The study [17] shows how blockchain technology can handle clinical trial consent, data, and results transparently and reliably. Such advancements in the conduct of clinical studies are critical to advancing medicine. Most clinical studies overextend their financial resources as well as their allocated time. Clinical and trial data exchange can hasten medical advancement in a competitive setting. Using various blockchain-based approaches like *Medrec*, *Patientory*, and *Medshare* can facilitate in data sharing for healthcare data management. The pharmaceutical industry's research sphere is also significant, covering everything from drug developers to medical device manufacturers to clinical trial results. Block RX provides a solution across this spectrum using advanced cryptography.

### 6.3.2 PRODUCT SUPPLY CHAIN DISTRIBUTION

There is a greater possibility of fraud when numerous mediators and merchants, as shown in Figure 6.5, are involved in pharmaceutical product distribution and its supply chain, which reduces the efficiency of the supply chain. Utilizing blockchain technology has been acknowledged for its capacity to prevent the dissemination of substandard drugs [18]. Products suspected of poor quality were withdrawn and investigated for origin. Ledger systems, chaincodes, and serialization facilitate the distribution of pharmaceuticals by assigning unique identifiers to individual products for monitoring, tracing, and allocation. Stringent regulations govern blockchain data to prevent attacks that could compromise security systems [19]. The Internet of Things (IoT) has enhanced the pharmaceutical distribution system's efficiency [20].

Table 6.1 depicts various blockchain cum IoT-based supply chain technological solutions for pharmaceutical industries.

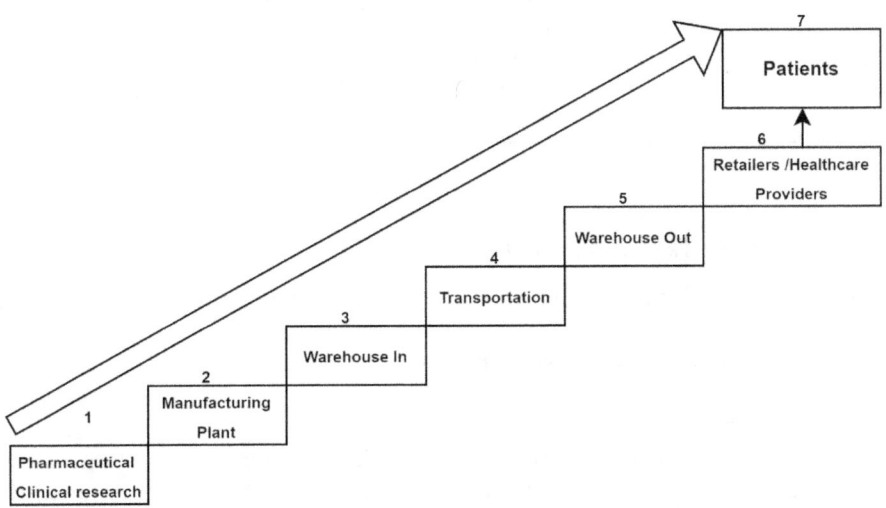

**FIGURE 6.5**  Pharmaceutical industry supply chain stakeholders.

**TABLE 6.1**
**Blockchain Initiatives for Pharmaceutical Supply Chain Solutions**

| Initiatives | Description |
| --- | --- |
| Ambrosus [21] | AMB-net is a blockchain-based IoT network specializing in agricultural and pharmaceutical supply chains. |
| MediLedger Project | It provides access-restricted blockchain applications for the pharmaceutical industry's compliance requirements with track-and-trace regulations. |
| Blockverify | It is a supply chain transparency and anticounterfeiting system for the prospective pharmaceutical industry and beyond applications. |
| Authentag | The pharmaceutical supply network can benefit from blockchain's distributed ledger technology. |
| Modsense T1 from Modum | The blockchain technology used to monitor supply chain ambient temperature and humidity. |
| IEEE Pharmacy Supply Chain Forum | A centralized forum where pharmaceutical industry and non-pharmaceutical industry stakeholders can investigate the potential of blockchain-based supply chain solutions. |
| DHL collaboration with Accenture | A prototype solution service employs blockchain technology to track pharmaceuticals' whole supply chain distribution. |
| Imperial Logistic Collaboration with One Network Enterprises [22] | The collaborative solution is based on One Blockchain technology from One Network Enterprises and improves supply chain security. |
| GFT collaboration with MYTIGATE | The result of this group effort is a proof-of-concept for using blockchain technology to track drugs. |
| EasySight Supply chain management/ Hejia | The EasySight blockchain-based solution keeps track of pharmaceuticals all along the supply chain. It gives complete transparency to trade records and helps smaller businesses get paid faster. |
| SAP | SAP has added blockchain technology to its advanced track and trace for pharmaceuticals (ATTP) to deal with supply chain problems in light of new regulations. |

### 6.3.3 Product Tracking and Tracing

Implementing tracking of the supply chain of drugs using blockchain involves several steps, as shown in Figure 6.6.

Below is an overview of the whole process:

1.  Define the scope
    Determine which drugs to track on the blockchain and what information to record for each drug. It includes information about the manufacturer, distributor, seller, and end consumer and details about the drug, such as the batch number, expiration date, and dosage.

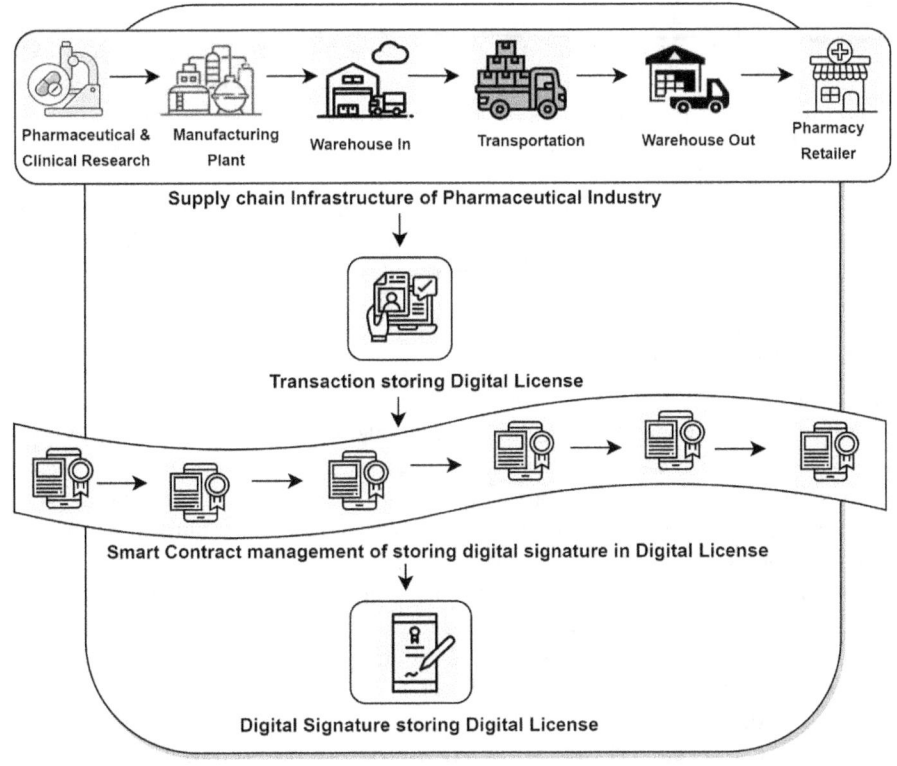

**FIGURE 6.6** Blockchain tracking of supply chain for drugs.

2.  Choose a blockchain platform
    Choose a blockchain platform that suits the supply chain's needs, considering scalability, security, and interoperability factors.
3.  Create a smart contract
    Develop a smart contract that will govern the tracking of the drugs on the blockchain. The smart contract should specify the rules for recording and verifying transactions and any penalties for non-compliance.
4.  Register participants
    Register all blockchain participants on the supply chain, including manufacturers, distributors, sellers, and end consumers. Each participant is assigned a unique identifier on the blockchain.
5.  Record transactions
    Record all transactions on the blockchain, including the transfer of ownership and the movement of drugs between participants. Each transaction should include details such as the date, time, location, and quantity of the drugs involved.
6.  Verify transactions
    Use the smart contract to verify transactions and ensure that they comply with the supply chain rules. It includes verifying the drugs' authenticity, checking for counterfeit products, and ensuring they are transported and stored correctly.

7.  Monitor the supply chain
    Monitor the supply chain in real time using blockchain analytics tools, and
    identify any anomalies or inconsistencies. It includes checking for changes in
    the ownership of drugs, discrepancies in the quantities of medicines carried,
    and unusual movement patterns.

By tracking the supply chain of drugs using blockchain, it is possible to create a
more secure and transparent supply chain, which can help reduce the risk of counterfeit medicines entering the market and improve patient safety.

### 6.3.4  DRUG COUNTERFEIT PREVENTION

After deaths were attributed to the distribution of substandard and counterfeit
medications, governments across the globe implemented trace and track systems
to monitor pharmacy supply chains [23]. Blockchain technology allows innovative
methods for tracing pharmaceuticals' origins [24]. It is simpler to monitor and track
the status of medication from the time they are manufactured until it reaches the
patient with blockchain [25, 26, 27, 28].

The pharmaceutical industry serializes its products and equips them with additional security features that purchasers can use to verify their authenticity and identify counterfeits. The blockchain system is secure due to transparent chaincodes for
transactions. Despite the dangers posed by counterfeit medicines continue to sell
due to a lack of trust in the system. Lives saved and safety increase if blockchain is
employed for quality control and counterfeit drug detection [29]. The Anti-Counterfeit
Medication System (ACMS) is one of many available instruments for combating
counterfeit pharmaceuticals. ACMS utilizes the Interplanetary File System (IPFS)
networks and the Ethereum blockchain in the following ways:

*   Establish ownership criteria for retail and non-retail medications to prevent
    duplication of medicines.
*   Construct Ethereum smart contracts for practical ACMS management by lever-
    aging IPFS networks and the Ethereum blockchain.
*   Implement the program for modest businesses.
*   Assess and investigate the proposed system [30].

The ACMS is an effective fraud prevention system. Before initiating a transaction,
customers must generate a chaincode. Transaction validation occurs at the end of
the chain after the confirming peers have validated the signature and endorsements
sent to the procuring services [31]. Several strategies have been proposed, including
implementing Ethereum blockchain and distributed ledger technologies for pharmaceutical supply chain management [24] to combat drug counterfeiting and improve
pharmaceutical traceability [32]. This system provides its users with numerous
advantages, including enhanced drug monitoring, user privacy, quality management,
non-repudiation, supply-side transparency, and demand-side management [25]. The
drug ledger is a blockchain-based, peer-to-peer architecture for scenario-based drug
control and traceability [9]. Hyperledger Fabric deploys and administers drug delivery

tracking systems for pharmaceutical businesses [33]. An Indian study quantified key blockchain characteristics such as transparency, immutability, and the capacity to record and trace medication data [31]. IoT devices can also verify the integrity of data sources and track the status of pharmaceutical products in real time and from anywhere. Blockchain technology facilitates data storage and sharing, ensuring that all records can be viewed and followed at any time [34]. Notably, the pharmaceutical industry in Italy has implemented a blockchain-based technological solution and serialization laws [19].

There are several ways to authenticate drugs using blockchain technology, including:

1.  *Digital fingerprinting:* Each drug is assigned a unique digital fingerprint and cryptographic hash of its characteristics, such as its batch number, expiry date, and other relevant information. This digital fingerprint recorded on the blockchain verifies the drug's authenticity at any point in the supply chain.
2.  *Quick Response (QR) codes:* Each drug is assigned a QR code that contains information about the drug, including its digital fingerprint. A smartphone or other device can scan the QR code, and the data can be verified against the blockchain to ensure the drug is genuine.
3.  *NFC tags:* Near-field communication (NFC) tags can be attached to each drug, which a smartphone or other device can read. These tags can authenticate the drug by verifying its digital fingerprint against the blockchain.
4.  *Tamper-evident packaging:* Tamper-evident packaging can provide physical evidence that the drug has not been tampered with. This packaging can contain a unique identifier recorded on the blockchain, which can verify the drug's authenticity.
5.  *Smart contracts:* Smart contracts enforce supply chain rules and ensure that only genuine trade of medicines is possible. For example, a smart contract can prevent the sale of drugs not authenticated on the blockchain.

### 6.3.5 SAFETY AND SECURITY

The architecture of conventional medication supply chain management does not permit the transmission of sensitive data. Usually, it is easy to modify, delete, or otherwise manipulate data. Numerous researchers have recently begun utilizing blockchain technology to transmit data across medicinal supply chains. The blockchain is a collection of time-stamped, immutable, sequential data units. A hashing algorithm created a digital imprint of the data for each block. This technology can safeguard the authenticity of medical records because each blockchain transaction is permanently documented and cannot be modified. In this section, we will provide a concise overview of how researchers have approached the various blockchain-based solutions proposed for assuring the network-wide integrity of medication data.

Bera et al. [11] investigated how blockchain technology aids pharmaceutical supply chains in meeting strict security requirements – a blockchain-based proof-of-concept design to comply with Drug Supply Chain Security Act rules. The Hyperledger Composer aids in modeling the various supply chain entities and access control rules before prototype implementation. As a result, blockchain applications could be developed and integrated with supply chain companies more rapidly. The Elliptic Curve Digital Signature Algorithm (ECDSA) employed by Sahoo et al. [27] and Zhang et al. [35] discussed blockchain technology's potential to prevent problems in managing the pharmaceutical supply chain securely and transparently. As part of a blockchain and IoT-driven network, wireless sensors and Global Positioning System (GPS) on the packages should record prescribed information such as temperature and location. Serial numbers and the manufacturer's fingerprints are imprinted on each medicine bottle. A consumer can obtain all the necessary information about a medicine by purchasing it and scanning its label with a smartphone.

Bocek [36] and Ying [37] proposed a blockchain technology-based system to supply prescription drugs to patients. This architecture can protect patient privacy by assigning a dynamic identity to each party and establishing an efficient authentication process. The design can also suit healthcare systems' computational and security requirements, such as authentication, user privacy, safe data sharing, visibility, and efficiency. The authors of [15] and [38] proposed a blockchain paradigm for pharmaceutical production and distribution because they were worried about data integrity. It verifies the authenticity of all incoming pharmaceuticals by monitoring the origins of their raw components and finished products. Under this system, the product manufacturer, pharmacy warehouse, and pharmacy have access to complete, reliable, and secure data about the source and quality of the drugs documented on the blockchain. Wu, and Long in [9] suggested a novel blockchain-based method for managing the drug supply chain, which aided in preserving drug data. The Hyperledger Fabric incorporated in the smart hospitals enables the secure storage and transfer of drug supply chain data.

There is an effective way for different hospital units to store and share electronic prescriptions and patient information. Electronic medication and patient health records have been given time-limited access using smart contracts to ensure data consistency. An access control policy is also specified to legitimize the proposed system's transaction requests further. The authors proposed a novel method of integrating and deploying a CouchDB for each node to prevent data duplication within the blockchain's underlying file system. Using IPFS as a data repository, researchers [39, 18] developed a patient-centric drug history recording system that can simplify collecting drug histories, recording them more accurately on the blockchain, and preventing data manipulation. QR code stores patient-encrypted prescription information. The database holds the hash value to prevent data manipulation and fraud. The blockchain-based PSCM proposed by Dwivedi, Amin, and Voll [19] employs smart contracts and consensus methods to facilitate secure data sharing in the pharmaceutical supply chain. With the aid of smart contract technology, the proposed system securely distributes cryptographic keys to all participants. When a new transaction occurs in the network, details get recorded in an immutable, time-stamped block that traces the origin and destination of a product throughout the entire supply chain. The authors of [40] and [41] created a custom smart contract for storing and querying pharmacogenomics data using the Ethereum blockchain and a multi-map, index-based strategy. Nodes use a mapping indexed by a unique identifier to efficiently store and retrieve each genomic annotation (a set of three genetic drugs and their outcomes). Each genomic note (a group of three genetic medicines and their results) is cataloged in the nodes with a unique identifier, enabling efficient storage and querying.

### 6.3.5.1 Data Governance

Blockchain and the IoT have facilitated data governance and supply chain progress [42]. Compliance with medical product regulations is yet another application of IoT [26]. The Drugledger system has been implemented in vaccine production to monitor transparency and regulations compliance [7]. Future technologies will also allow for the distribution of vaccines [28].

### 6.3.5.2 Data Quality

IoT monitoring medical product temperatures during distribution can also alert relevant parties to take necessary corrective action [43]. However, another IoT system verifies the authenticity of pharmaceuticals [23]. Due to its data administration skills (transparency and immutability), it has been used in a few studies to identify low-quality medications [44]. Another system encourages open communication between the numerous recall process participants [27]. Recall management can increase productivity and accountability in additional ways, including safeguarding the confidentiality of product recall data. Stochastic models protect pharmaceutical products from larceny and temperature fluctuations in several markets where open administration and counterfeit protection are absent [45].

### 6.3.5.3 Pharmaceutical Sales Volume

The use of blockchain technology (in this case, Hyperledger Fabric) to control the flow of pharmaceuticals is so novel that its long-term utility is difficult to predict [34]. The RxCoin smart contract was implemented to combat the Opioid crisis – based on a digital currency used to describe medications using Ethereum blockchain technology [39] digitally. RxCoin's success paves the way for a blockchain-based prescription library. RxCoin contracts provide no evidence of a prescription drug monitoring program (PDMP) protected health information implementation on the blockchain, and HIPAA protection regulations are not considered for stored PHI. Interfaces for RxCoin are currently under development. Researchers are assiduously working to cover a market void by enhancing the security of RxCoin contracts and employing digital signatures as system members' identities.

### 6.3.6 Prescription Management

Blockchain technology can revolutionize prescription management by providing a secure, transparent, and tamper-proof system. Here are some potential applications of blockchain in prescription management:

1. *Prescription tracking:* Blockchain builds an immutable record of prescriptions, including the prescribing doctor, the patient, the medication prescribed, and

the dispensing pharmacy. It helps to prevent duplicate prescriptions and track the dispensing of controlled substances, reducing the risk of prescription drug abuse.

2. *Prescription filling:* Blockchain creates a secure and transparent record of prescription filling, reducing the risk of prescription errors and ensuring the patient receives the correct medication.
3. *Supply chain management:* Blockchain tracks the medication supply chain from the manufacturer to the pharmacy. It helps to prevent counterfeit drugs and ensure the quality of the medication.
4. *Patient-controlled data:* Blockchain gives patients more control over their medical data, including their prescription history. Patients can use a private key to control who can access their prescription history, helping to protect their privacy.
5. *Automated prescription refills:* Blockchain automates prescription refills, reducing the need for manual verification and the risk of errors.
6. *Interoperability:* Blockchain can create a more interoperable healthcare system, allowing healthcare providers to access and share prescription data more efficiently.

Prescription management is essential for healthcare administration efficiency. Recently, prescription drug abuse has become more prevalent, contributing to crises such as opioids [46] BlockMedx uses an Ethereum-based platform to administer prescription procedures and transactions recorded on a blockchain securely. It is one of the numerous blockchain-based approaches to overcoming the obstacles to effective prescription management. Before dispensing medication to a patient, a pharmacist can verify a doctor's prescription using blockchain technology. This method makes it easier to manage controlled substances, such as narcotics. Project Heisenberg is another example of software for monitoring medication use based on Ethereum smart contracts. Customers, physicians, and pharmacies each have their portal within the system following their respective roles in the prescription process. ScriptDrop has facilitated the simplification of patient pharmacy distribution. Patients no longer need to remember to pick up their prescriptions because they are delivered to their homes. Using digital tools, they also monitor medication intake (adherence). Using blockchain technology, ScriptDrop tracks compliance and transportation data. ScalaMed is a blockchain-based, patient-centric system for managing healthcare compliance and monitoring prescriptions (including historical ones). This chapter describes the proposed solution as an electronic "prescription inbox" that will eliminate the issue of incorrect medication administration. Current hospital treatment may interact negatively with the new medication the family physician prescribes. These patient-accessible data points are difficult to centralize under the current system, which may have unintended consequences. ScalaMed's solution centralizes patients' current and historical prescription histories, reducing the likelihood of adverse drug interactions. Despite investigating numerous blockchain-based solutions for prescription management, some conventional, centralized systems may also provide a solution for markets with few participants. A simple solution, such as a single sign-on into a shared database to view a patient's medication history, could be anticipated through close collaboration and a collaborative tool by the clinician and the pharmacy (e.g., due to regulatory enforcement).

### 6.3.7  BILLING CLAIMS MANAGEMENT

Blockchain technology improves billing and claim management in healthcare in the following ways:

1. *Streamlining the claims process:* One of the main advantages of blockchain technology is that it allows for secure and efficient data transfer. In healthcare, this can streamline the claims process by reducing the time and costs associated with claims processing. For example, a blockchain-based system could automate the claims process, making it faster and more efficient.
2. *Improved transparency and accuracy:* Blockchain technology allows for transparent and auditable records, which can help to reduce errors and fraudulent claims. A blockchain-based system can ensure that all claims are legitimate and accurate by creating a secure and transparent record of all transactions.
3. *Secure storage of sensitive information:* The healthcare industry deals with sensitive patient information, such as medical records and billing information. Blockchain technology can provide a secure and tamper-proof platform for storing information to protect it from cyber threats and data breaches.
4. *Reduced administrative costs:* By automating the claims process and reducing errors, blockchain technology can help to reduce administrative costs associated with claims processing. It helps to make healthcare more affordable for patients and reduces costs for healthcare providers.
5. *Faster payment processing:* Blockchain technology can enable speedier payment processing by automating the claims process and reducing the time it takes for claims to be processed and paid. It helps improve healthcare providers' cash flow and reduce the financial burden on patients.

The financing of healthcare is an integral component of the healthcare system. The use of blockchain technology has the potential to enhance the efficiency of the healthcare financing system, especially in terms of confidence and transparency. Since blockchain's distributed ledger technology incorporates trust, it can facilitate direct links between patients (who submit claims) and bearers (who clear the claim). The process of determining premiums is an ideal place to implement smart contracts. Information on present health, medication use, and lifestyle is linked via blockchain to fluctuating premiums via smart contracts. When multiple parties or intermediaries are processing a claim, the ultimate customer may be subject to repeated fees and audits.

In light of these gaps, proposals to use blockchain for invoicing, claims management, and other financial aspects of healthcare delivery came under consideration. Gem uses blockchain technology to facilitate healthcare claims processing and service delivery. It integrates patients, healthcare providers, and payers into a single ecosystem to enhance real-time patient data and streamline the health claims process.

Change Healthcare can manage its claims and revenues on the blockchain using the Hyperledger Fabric 19 framework. HSBlox has released the Redbox (TM) and CuraBlox (TM) products for claims administration atop their blockchain platform Simplified Exchange and Transparency for Users (SETU). The company employs machine learning to enable automated decision-making on top of claims, such as the detection of duplicate claims or the identification of patterns in the denial of claims, thereby further enhancing the transparency and trust provided by the blockchain. Dockchain, created by Pokidot [47], enables the processing of financial data on the blockchain in a medical context using tools such as smart contracts. Insurers such as Humana and United Healthcare are among the numerous healthcare organizations working on a trial program [48] that will use blockchain to store and disseminate curated healthcare provider data. This function prevents many errors and unnecessary stages in the insurance claim process. Table 6.2 summarizes efforts to implement blockchain-enabled healthcare invoicing and claim administration solutions.

**TABLE 6.2**
**Blockchain-Based Solutions for Healthcare Billing and Claim Management**

| Initiatives | Descriptions |
|---|---|
| Change Healthcare [49] | HyperLedger version 1.0-based claims and income control platform. |
| Gem | Efficiency in hospital claims handling using Ethereum-based blockchain technology. |
| HSBlox | Simplified Exchange and Transparency for Users (SETU), a blockchain-based tool, addresses claims of management issues. |
| Solve.Care | Solve.Care, the blockchain-based solution, provides decentralized health management, particularly preventing abuse and fraud in various health benefits programs. |
| Pokitdok | Dockchain is a cryptocurrency designed to facilitate using smart contracts to manage financial data in a healthcare setting. |
| Smartillions | Smartillions's approach relied on a blockchain-based claims management system, with payments coming from a pension fund and the opportunity for all provider transactions to settle in cryptocurrency. |
| Health Nautica collaboration with Factom | The partnership uses a blockchain-based solution to ensure that all financial transactions related to medical billing are safe. |
| Robomed Network | By linking payment for a medical procedure to the expected clinical result, Robomed's blockchain-based solution incentivizes providers to get the treatment right the first time. |
| Quantum Medical Transport collaboration with River Oaks Billing Associates | The collaboration is developing a blockchain-based system for managing claims and other data records. |

### 6.3.8 Challenges and Future Research Opportunities of Blockchain in the Pharma Sector

Blockchain technology in the pharmaceutical business resolves data reliability, data dredging, and other problems in clinical trials, as well as patient healthcare-related issues [50, 51]. The following are some of the most serious issues raised by using blockchain technology in the pharmaceutical business [52, 53]. Figure 6.7 outlines blockchain technology's challenges and issues in the pharmaceutical business.

#### 6.3.8.1 Security

Security concerns surround the use of blockchain technology in the pharmaceutical industry. Since blockchain applications require the Internet for access, they may be vulnerable to cyberattacks like denial-of-service, theft, and spying, which can upset blockchain services. The 51% attack, also called the majority attack, is one way to break into a cryptocurrency system [54]. This attack makes it possible to reject transactions and double-spend or spend the same coins more than once. This attack is more brutal to pull off if there are other cryptocurrencies with communities of miners besides Bitcoin. So, security is critical in protecting against these kinds of attacks. Blockchain technology and its models across multiple platforms are more vulnerable [53].

#### 6.3.8.2 Integration

When blockchain solutions are combined with other distributed applications, making new apps for the pharmaceutical industry will be possible. The security and compatibility issues arising due to integration are problematic. Proof of this challenge is that blockchain-based solutions in the pharmaceutical business require compatibility with various platforms and operating systems [55]. The integration process is complicated because integrated pharmaceutical organizations typically use multiple environments, methodologies, and computer languages. Future pharmaceutical industrial sectors will benefit from safe, consistent, reliable, and readily available integration [56, 57].

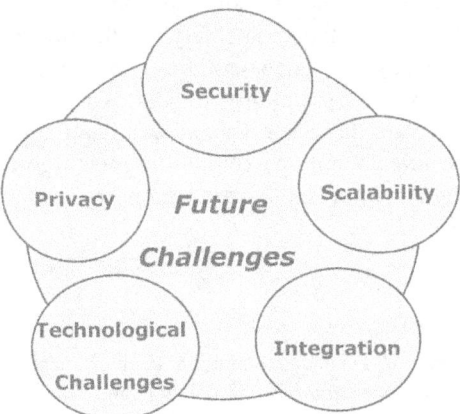

**FIGURE 6.7** Blockchain technology challenges in the pharmaceutical business.

### 6.3.8.3   Scalability

Since modern pharmaceutical industries rely on blockchain technology which produces a high volume of transactions for linking and processing, healthcare system efficiency may suffer [58]. The pharmaceutical industry is susceptible to this issue because the process slows down as the blockchain grows. More participants in a deal mean more potential problems with scaling. Due to their applications, these issues lead to operational and financial challenges in the pharmaceutical industries [57, 59].

### 6.3.8.4   Privacy

The changes or contributions made to a public blockchain by some pharmaceutical industry members can be verified and observed by those with access to the blockchain. Nonetheless, there is a binding agreement on authorization rules that all parties must follow. Because of the public nature of the blockchain, it can be challenging to maintain tracks on all the stakeholders in the pharmaceutical industry who have access to the complete transition. Further, since different applications call for different kinds of blockchains, the operation and management in the private blockchain migration process are handled by a single entry subject to its rules and regulations [60]. Private pharmaceutical businesses' high levels of security and privacy are required of their participants [57].

### 6.3.8.5   Technological Challenges

The need for more technical and technological knowledge of blockchain technology is a severe challenge for the adoption of blockchain technology in pharmaceutical industries [59]. It requires a certain degree of literacy in the complex and straightforward blockchain. Thus, efforts are necessary to improve user experience and develop blockchain protocols [61, 62, 63, 64, 65, 66].

## CONCLUSION

This study underscores the transformative potential of blockchain technology in the pharmaceutical industry, heralding advancements in combating counterfeiting, streamlining distribution, and enhancing safety. While blockchain offers promising solutions for tracking, data governance, and drug monitoring, significant challenges such as security, privacy, and scalability remain. The path forward necessitates focused research to navigate these hurdles and validate the practical application of blockchain in pharmaceuticals. Future explorations must rigorously evaluate its real-world efficacy, marking a pivotal step toward a more transparent and secure pharmaceutical paradigm.

## REFERENCES

1.   Soundarya K, Pandey P, Dhanalakshmi R. A counterfeit solution for pharma supply chain. *EAI Endorsed Trans Cloud Syst*. 2018; 3(11):e5.
2.   Rogaway P, Shrimpton T. Cryptographic hash-function basics: Definitions, implications, and separations for preimage resistance, second-preimage resistance, and collizion

resistance. In: *International Workshop on Fast Software Encryption*. Springer, pp. 371–88.

3. Nakamoto S. *Bitcoin: A Peer-to-Peer Electronic Cash System. Decentralized Bus. Rev.* 2008.

4. Reshi IA, Sholla S. Challenges for security in iot, emerging solutions, and research directions. *Int J Comp Digital Syst.* 2022; 12(1): 1231–41.

5. Haq I, Muselemu O. Blockchain technology in the pharmaceutical industry to prevent counterfeit drugs. *Int J Comput Appl.* 180(25):8–12.

6. MacDonald L. Trading globally in Austrian history: Vereinigte bühnen wien. In: *The Palgrave Handbook of Musical Theatre Producers*. pp. 343–9. Springer. 2017.

7. Sinclair D, Shahriar H, Zhang C. Security requirement prototyping with hyper ledger composer for drug supply chain: A blockchain application. In: *Proceedings of the 3rd International Conference on Cryptography, Security and Privacy – ICCSP '19*. Kuala Lumpur. Malaysia.

8. Plotnikov V, Kuznetsova V. The prospects for the use of digital technology 'blockchain' in the pharmaceutical market. In: *MATEC Web Conference*. 2018; 193: 02029. DOI

9. Huang Y, Wu J, Long C. Drugledger: A practical blockchain system for drug traceability and regulation. In: *2018 IEEE International Conference On Internet of Things(iThings) and IEEE Green Computing and Communications (GreenCom) and IEEEcyber, Physical and Social Computing (CPSCom) and IEEE Smart Data*. SmartData. Piscataway: IEEE; pp. 1137–44. 2018.

10. Garankina RY, Zakharochkina ER, Samoshchenkova IF, Lebedeva NY, AV.2018 L. Blockchain technology and its use in the area of circulation of pharmaceuticals. *J. Pharm. Sci.* 10(11):2715–7.

11. Bera B, Saha S, Das AK, Kumar N, Lorenz P, Alazab M. 2020. Blockchain-envisioned secure data delivery and collection scheme for 5G-Based IoT-enabled Internet of drones environment. *IEEE Trans Veh Technol.* 69(8):9097–111.

12. Bhardwaj A, Shah SBH, Shankar A, Alazab M, Kumar M, Gadekallu TRA. Penetration testing framework for smart contract blockchain. *Peer-to-Peer Netw Appl.* 14:2635–50.

13. Kumar R, Tripathi R, Marchang N, Srivastava G, Gadekallu TR, Xiong NN. A secured distributed detection system based on IPFS and blockchain for industrial image and video data security. *J Parallel Distrib Comput.* 152:128–43.

14. Alshahrani W, Alshahrani R. Assessment of blockchain technology application in the improvement of pharmaceutical industry. In: *2021 International Conference of Women in Data Science at Taif University (WiDSTaif)*. pp. 1–5.

15. Makarov AM, Pisarenko EA. Blockchain technology in the production and supply of pharmaceutical products. In: *Proceedings of the International Scientific and Practical Conference on Digital Economy (ISCDE)*. 2019.

16. Brooks RR, Wang, KC, Yu L, Oakley J, Skjellum A, Obeid JS and Lenert L, Worley C. Scrybe: A blockchain ledger for clinical trials. In: *IEEE Blockchain in Clinical Trials Forum: Whiteboard challenge winner* (pp. 1–2). IEEE New Jersey. 2018.

17. Ravaud MB. *Blockchain Technology for Improving Clinical Research Quality*. Trials.

18. Hulea M, Rosu O, Miron R, Astilean A. Pharmaceutical cold chain management platform based on a distributed ledger. In: *2018 IEEE International Conference on Automation, Quality and Testing, Robotics (AQTR)*. Piscataway: IEEE.

19. Dwivedi SK, Amin R, Vollala S. Blockchain based secured information sharing protocol in supply chain management system with key distribution mechanism. *J Inf Secur Appl.* 2020; 54:102554.

20. Botcha KM, Chakravarthy VSSS, Anurag. Enhancing traceability in pharmaceutical supply chain using internet of things (iot) and blockchain. In: *Proceedings – 2019 IEEE*

*International Conference On Intelligent Systems and Green Technology, ICISGT.* 2019; 201945–48.

21.  Craib R, Bradway G, Dunn X. Ambrosus white paper [Internet]. Available from: https://ambrosus.com/assets/en/Ambrosus-White-Paper.pdf

22.  No title [Internet]. Available from: www.onenetwork.com/2012/01/one-network-ente rprises-and-imperial-logistics-partner-to-achieve-supply-chain-excellence-visibility-and-flexibility-creates-next-generation-supply-chain/

23.  Sylim P, Liu F, Marcelo A, Fontelo P. Blockchain technology for detecting falsified and substandard drugs in distribution: pharmaceutical supply chain intervention. *JMIR Res Protoc.* 2018;7(9): e10163.

24.  Raj R, Rai N, Agarwal S. Anticounterfeiting in pharmaceutical supply chain by establizhing proof of ownership. In: *TENCON 2019–2019 IEEE Region 10 Conference (TENCON).* Piscataway: IEEE.

25.  Jangir S, Muzumdar A, Jaiswal A, Modi CN, Chandel S, Vyjayanthi C. A novel frame-work for pharmaceutical supply chain management using distributed ledger and smart contracts. *2019 10th International Conference on Computing, Communication and Networking Technologies (ICCCNT) (pp. 1–7).* 2019. IEEE.

26.  Ahmadi V, Benjelloun S, El Kik M, Sharma T, Chi H, Zhou W. *Drug Governance:IoT-Based Blockchain Implementation in the Pharmaceutical Supply Chain.* IEEE. 2020.

27.  Sahoo M, Singhar SS, Nayak B, Mohanta BK. A blockchain based framework secured by ECDSA to curb drug counterfeiting. In: *2019 10th International Conference on Computing, Communication and Networking Technologies (ICCCNT).* pp. 1–6.

28.  Surjandy S, Widjaja M, Oktriono K, Fernando E. Benefit and challenge of blockchain technology in pharmaceutical supply chain management. *Int J Recent Technol Eng.* 8(4):8309–13.

29.  Adsul KB, Kosbatwar SP, Kajal M, Adsul B. A novel approach for traceability & detection of counterfeit medicines through blockchain [Internet]. University of Manchester: EasyChair, Available from: https://easychair.org/publications/prepr int/QJNf

30.  Saxena N, Thomas I, Gope P, Burnap P, Kumar N. Pharmacrypt: blockchain for crit-ical pharmaceutical industry to counterfeit drugs. *Computer (Long Beach Calif).* 53(7):29–44.

31.  Kumar A, Choudhary D, Raju MS, Chaudhary DK, Sagar RK. Combating counter-feit drugs: A quantitative analysis on cracking down the fake drug industry by using blockchain technology. In: *2019 9th international conference on cloud computing, data science \& engineering (Confluence)* (pp. 174–178). IEEE. 2019.

32.  Pham HL, Tran TH, Nakashima Y. *Practical Anti-Counterfeit Medicine Management System Based on Blockchain Technology.*

33.  Din ZU, Pervez L, Amir A, Abbas M, Khan I, Iqbal Z, et al. Parasitic infections, malnutrition and anemia among preschool children living in rural areas of Peshawar, Pakistan. *Nutr Hosp [Internet].* 35(5):1145–52. Available from: https://doi.org/10.20960/nh.1685

34.  Shi J, Yi D, Kuang J. Pharmaceutical supply chain management system with integration of IoT and blockchain technology. In: *Smart Blockchain.* Cham: Springer International Publizhing, pp. 97–108.

35.  Zhang J, Shi X, Xie J, Ma H, King I, Yeung D-Y. *Gaan: Gated Attention Networks for Learning on Large and Spatiotemporal Graphs.* arXiv Prepr arXiv180307294. 2018.

36.  Bocek T, Rodrigues BB, Strasser T, Stiller B. Blockchains everywhere – a use case of blockchains in the pharma supply-chain. In: *2017 IFIP/IEEE Symposium on Integrated Network and Service Management.* IM. Piscataway: IEEE, pp. 772–7.

37.  Ying B, Sun W, Mohsen NR, Nayak A. A secure blockchain-based prescription drug supply in healthcare systems. In: *2019 International Conference on Smart Applications, Communications and Networking (SmartNets)*. Sharm El Sheik, Egypt, pp. 1–6. DOI

38.  Fernando E. Success factor of implementation blockchain technology in pharmaceutical industry: a literature review. In: *2019 6th international conference on information technology, computer and electrical engineering (ICITACEE)* (pp. 1–5). IEEE. 2019.

39.  Kim JW, Lee AR, Kim MG, Kim IK, Lee EJ. Patient-centric medication history recording system using blockchain. In: *2019 IEEE International Conference on Bioinformatics and Biomedicine (BIBM)*. San Diego, CA, USA.

40.  Gürsoy G, Brannon CM, Gerstein M. Using Ethereum blockchain to store and query pharmacogenomics data via smart contracts. *BMC Med Genomics*. 2020; 13(1): 1–11.

41.  Archa B, Alangot B, Achuthan K. Trace and track: Enhanced pharma supply chain infrastructure to prevent fraud. Kumar N, Thakre A, editors. *Soc Informatics Telecommun Eng*. 218:189–95.

42.  Jalali S, Wohlin C. Systematic literature studies: Database searches vs. backward snowballing. In: *Proceedings of the ACM- IEEE International Symposium on Empirical Software Engineering and Measurement – ESEM '12*. Piscataway: IEEE.

43.  Tseng JH, Liao YC, Chong B, Liao SW. Governance on the drug supply chain via gcoin blockchain. *Int J Environ Res Public Health*. 15(6).

44.  Badhotiya GK, Sharma VP, Prakash S, Kalluri V, Singh R. Investigation and assessment of blockchain technology adoption in the pharmaceutical supply chain. *Mater Today Proc*. 2021; 46:10776–80.

45.  Liberati A, Altman DG, Tetzlaff J, Mulrow C, Gøtzsche PC, Ioannidis JP, et al. The prisma statement for re porting systematic reviews and meta-analyses of studies that evaluate health care interventions: Explanation and elaboration. *Ann Intern Med*. 2009; 151(4): W–65.

46.  Skolnick P. The opioid epidemic: Crisis and solutions. *Annu Rev Pharmacol Toxicol*. 2018; 58: 143–59.

47.  Smith BW. In: *Dokchain: Intelligent Automation In Healthcare [Internet]*. Available from: https://pokitdok.com/wp-content/themes/pokitdok2017/dokchain/static/data/DokChainWhitepaper20170926Draft.pdf

48.  *Optum, United Healthcare, Humana, Others Launch Blockchain Pilot | Healthcare IT News [Internet]*. [cited 2023 Apr 2]. Available from: www.healthcareitnews.com/news/optum-unitedhealthcare-humana-others-launch-blockchain-pilot

49.  Healthcare C. *Retrieved from Better Healthcare [Internet]*. Available from: www.changehealthcare.com/

50.  Seliem M, Elgazzar K. BIoMT: Blockchain for the internet of medical things. In: *2019 IEEE International Black Sea Conference on Communications and Networking* (pp. 1–4). IEEE. 2019.

51.  Albanese G, Calbimonte JP, Schumacher M, Calvaresi D. Dynamic consent management for clinical trials via private blockchain technology. *J Ambient Intell Humaniz Comput*. 2020; 11: 4909–26.

52.  Long Y, Wu M, Liu Y, Fang Y, Kwoh CK, Chen J, et al. Pre-training graph neural networks for link prediction in biomedical networks. *Bioinformatics*. 2022 Apr;38(8):2254–62.

53.  Opportunities for use of blockchain technology in medicine. *Applied Health Econmics and. Health Policy (New York)*. 16(5):583–90.

54.  Jovic A, Jozic K, Kukolja D, Friganovic K, Cifrek M. Challenges in designing software architectures for web-based biomedical signal analysis. *Med Big Data IoMT*. 2018; 81–111.

55. Kleinaki AS, Mytis-Gkometh P, Drosatos G, Efraimidis PS, Kaldoudi E. A blockchain-based notarization service for biomedical knowledge retrieval. *Comput Struct Biotechnol J.* 16:288–97.

56. Xie J, Tang H, Huang T, Yu FR, Xie R, Liu J, Liu Y. A survey of blockchain technology applied to smart cities: Research issues and challenges. *IEEE Commun Surv Tutorials.* 2019; 21(3): 2794–2830.

57. Al-Jaroodi J, Mohamed N. Blockchain in industries: A survey. *IEEE Access.* 7:36500–15.

58. Hussein AF, ALZubaidi AK, Habash QA, Jaber MM. An adaptive biomedical data managing scheme based on the blockchain technique. *Appl Sci.* 2019; 9(12): 2494. 2019

59. Saberi S, Kouhizadeh M, Sarkis J, Shen L. Blockchain technology and its relationships to sustainable supply chain management. *Int J Prod.* 2019; 57(7):2117–2135.

60. Siyal AA, Junejo AZ, Zawish M, Ahmed K, Khalil A, Soursou G. Applications of blockchain technology in medicine and healthcare. *Challenges Futur Perspect.* 3(1):3.

61. Chang MB, Ullman T, Torralba A, Tenenbaum JB. *A Compositional Object-Based Approach To Learning Physical Dynamics.* arXiv Prepr arXiv161200341. 2016.

62. Moin S, Karim A, Safdar Z, Safdar K, Ahmed E, Imran M. Securing IoTs in distributed blockchain: Analysis, requirements and open issues. *Futur Gener Comput Syst.* 100:325–43.

63. Sarkar S, Saha K, Namasudra S, Roy P. An efficient and time saving web service based android application. *SSRG Int J Comput Sci Eng (SSRG-IJCSE).* 2(8):18–21.

64. Namasudra S, Roy P, Balamurugan B, Vijayakumar P. Data accessing based on the popularity value for cloud computing. In: *Proceedings of the International Conference on Innovations in Information, Embedded and Communications Systems (ICIIECS).* India: IEEE, Coimbatore; pp. 109–13.

65. Namasudra S, Deka GC. *Advances of DNA Computing in Cryptography.* 2018. CRC Press. Taylor Fr ISBN. 9780815385.

66. Sarkar S, Parmar A, Singh A. An exploratory study of cannabis use pattern and treatment seeking in patients attending an addiction treatment facility. *Indian J Psychiatry.* 2020 Mar;62(2):145–51.

# 7 Edge Intelligence Decentralized Blockchain-Based Internet of Things (B-IoT) for Sustainable Healthcare

*Sachin Sharma,[1] Ranu Tyagi,[1] and Seshadri Mohan[2]*

[1] Department of Computer Science and Engineering, Graphic Era (Deemed to be University), Dehradun, Uttarakhand, India

[2]Systems Engineering Department, University of Arkansas at Little Rock, Little Rock, Arkansas, United States

## 7.1 INTRODUCTION

Decentralized Edge Intelligence—a cutting-edge technology called blockchain-based Internet of Things (B-IoT) for sustainable healthcare—integrates the Internet of Things (IoT), blockchain, and edge computing to improve healthcare delivery. It entails using edge computing devices online and gathering information from several sensors and gadgets, including wearables and medical gear. Artificial intelligence and machine learning algorithms are then used to analyze the obtained data to produce insights that can be used to enhance healthcare results. Only authorized personnel can access the data because it is safe and impervious to tampering, thanks to blockchain technology. A decentralized B-IoT system enables healthcare professionals to safely communicate patient data across numerous platforms and devices, improving collaboration and care coordination. By enabling remote patient monitoring, individualized treatment regimens, and early diagnosis of health issues, this technology has the potential to transform the way healthcare is delivered completely.

Additionally, by increasing the effectiveness of healthcare delivery and lowering the need for hospitalization, B-IoT can aid in lowering healthcare expenditures. Minimizing resource consumption and waste can also contribute to advancing sustainability by lowering the environmental effect of healthcare. B-IoT is a promising

DOI: 10.1201/9781003407096-7

**109**

technology that could revolutionize healthcare delivery and encourage environmentally friendly medical procedures. Its application may lead to better patient outcomes, increased healthcare delivery efficiency, and a more resilient healthcare system.

## 7.2 LITERATURE REVIEW

The authors of [1] provide an in-depth analysis of blockchain for the Internet of Things (B-IoT) and explore the lessons learned from this novel paradigm. Specifically, they first briefly describe IoT and its difficulties. They talked about the problems of adopting blockchain for IoT in the fifth generation and beyond and the industrial uses of B-IoT. In [2], the authors looked into how IoT and blockchain technology could work together and presented a detailed analysis of the blockchain-enabled IoT and industrial IoT systems. Data storage and management techniques, big data and cloud computing techniques (finance and data auditing), and industrial sectors (supply chain, energy, and healthcare sector) are the categories under which state-of-the-art research is divided. The authors of [3] propose a blockchain–Internet of Things paradigm that addresses these problems by having a biosensor measure and gather real-time data about a patient's medical status and store it in the blockchain. This enables speedy reporting and tamper-proof data storage. The ultimate medical bill and insurance coverage can be estimated by implementing a smart contract. As a result, there would be no need for third-party service providers, and the system would be open. IoT healthcare procedures are examined in this context in [4], and a thorough explanation is provided. Additionally, it starts a thorough assessment of IoT healthcare services and applications. IoT surrounds healthcare and provides extensive insights into IoT healthcare security, including its requirements, problems, and privacy concerns. The authors in [5] discuss the various software-based data traffic management approaches, data centers, and energy-efficient hardware design concepts as Green Internet of Things enablers. In order to achieve the best power budgeting, energy models of IoT devices are described in terms of data transfer, actuation process, static power dissipation, and generated power. Tornado, a high-performance blockchain system based on a space-structured ledger and associated algorithms, was discussed in [6] to enable blockchain in the IoT. To help the network scale, they first create a space-structured chain architecture with new data structures. A unique consensus technique called collaborative proof of work is designed to address the massive heterogeneity of IoT. To increase the resource efficiency of IoT devices, they suggested the space-structured greedy heaviest-observed subtree (S2 GHOST) protocol. In S2 GHOST, a dynamic weight assignment technique also helps to reflect the reliability of the data and the devices. Numerous tests show that Tornado can process 3464.76 transactions per second at its maximum throughput. Propagation delay and resource efficiency have been optimized by 68.14% and 30.56%, respectively. In [7], the authors give a summary of the body of research on blockchain for IoT and lay out a road map of research issues that must be resolved before blockchain technology can be used in the IoT. In [8], a consensus algorithm utilizing a practical Byzantine fault-tolerant mechanism is used to improve blockchain-based transactions (BCTs). The suggested algorithm can increase the efficiency of data registration, transaction volume, and privacy protection of BCT. It is concluded that the dynamic game strategy of node cooperation can be used to prevent local domination. Furthermore, the state of the unknown node can be roughly determined by reporting the node's global reputation rating. In [9] the authors provided incentives for IoT devices to participate in the development of the wireless blockchain network and presented a multi-dimensional contract to maximize the utility of the blockchain while addressing the challenges of moral hazard and adverse selection. The proposed contract specifically examines device connectivity from the standpoint of complex network theory and considers IoT devices' hash power and communication costs. Through simulations with different network sizes and average link probabilities, they looked at the wireless blockchain network's energy consumption and block confirmation probability. The suggested contract mechanism is practical, generates 35% more utility than previous alternatives, and enhances utility by four times compared to the original proof-of-work (-based incentive mechanism, according to numerical results. In order to provide end-to-end security in resource-constrained IoT networks, the authors suggested a unique trust-based access control mechanism for edge-IoT networks using blockchain technology (called TABI) [10]. The TABI method uses access control and trust evaluation procedures to lessen the impact of malicious IoT users and devices.

## 7.3  OVERVIEW OF SUSTAINABLE HEALTHCARE

The practice of providing healthcare in a way that maximizes patient well-being while reducing negative environmental effects and preserving resource efficiency is known as sustainable healthcare. It consists of a wide range of practices and initiatives to enhance health outcomes, reduce waste, and promote social responsibility. Sustainable healthcare demands a multifaceted approach considering the social, economic, and environmental factors that impact health. It emphasizes the need for early identification, prevention, and evidence-based methods and equipment to improve health outcomes. Additionally, it prioritizes limiting healthcare waste and utilizing resources like power, water, and medical supplies as efficiently as possible.

Several examples of environmentally friendly medical operations are presented in Figure 7.1.

1. **Energy-efficient healthcare facilities:** Developing energy-efficient healthcare facilities can reduce energy use and greenhouse gas emissions.
2. **Trash reduction:** Reducing trash through programs like composting and recycling helps lessen the environmental effect of the healthcare industry.
3. **Sustainable procurement:** Buying products and services that are socially and environmentally responsible can help sustain local economies and communities.
4. **Remote patient monitoring:** Hospitalizations can be avoided, and patient outcomes can be enhanced using remote patient monitoring technologies.
5. **Health education:** Health education can encourage healthy habits and prevent chronic diseases by being given to patients and communities.

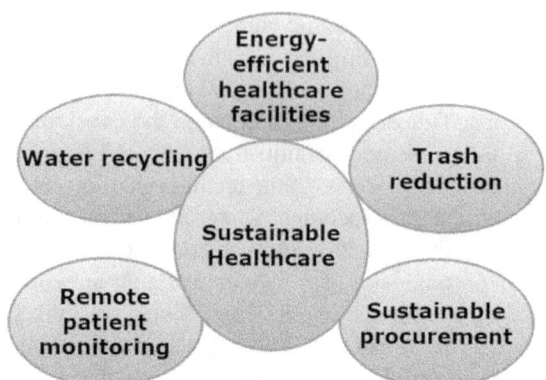

**FIGURE 7.1**   Overview of sustainable healthcare.

Sustainable healthcare is essential to encouraging environmental and social responsibility while making sure that healthcare services are available, affordable, and effective for everyone. Healthcare providers may enhance patient outcomes, cut costs, and build a more sustainable healthcare system by using sustainable healthcare practices.

## 7.4   TECHNOLOGIES IN SUSTAINABLE HEALTHCARE

For sustainable healthcare, a number of technologies are being used or developed. Here are a few instances (Figure 7.2).

1. **Electronic health records (EHRs):** EHRs are digital patient health records that healthcare professionals can access and share securely. With this technology, paper records are no longer necessary, less waste is produced, and healthcare providers can better coordinate patient care.
2. **Telemedicine:** Telemedicine uses technology, such as video conferencing, to deliver healthcare remotely. By lowering travel requirements, this technology can lower carbon emissions and provide access to healthcare services in remote places.
3. **Wearable medical equipment:** Wearable medical equipment, such as fitness trackers and smartwatches, can track a patient's health and give medical professionals real-time information. By enabling patients to manage their health better and preventing chronic diseases, this technology can lessen the need for hospitalization.
4. **Energy-efficient healthcare facilities:** Solar panels, LED lighting, and other energy-saving technology can be used to design energy-efficient healthcare facilities. Energy utilization and greenhouse gas emissions may be decreased as a result.
5. **Blockchain:** Blockchain technology can be used to store and distribute patient health information safely, allowing medical professionals to better coordinate

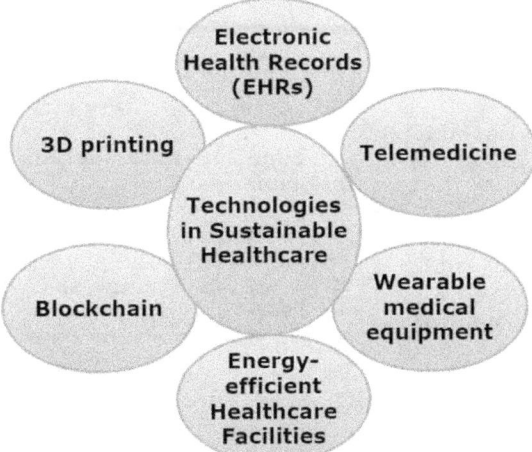

**FIGURE 7.2** Technologies in sustainable healthcare.

patient care and lowering the risk of data breaches. Additionally, this technology can allow patients autonomy over their health information management and sharing.

6. **3D printing:** By using 3D printing to make prosthetic limbs and medical implants, waste may be reduced, and patients can get personalized medical equipment.

## 7.5 PROPOSED METHODOLOGY

Edge Intelligence's decentralized blockchain-based Internet of Things (B-IoT) for sustainable healthcare has various stages that may be broken down (Figure 7.3):

1. **Data gathering:** The initial step entails gathering information from various sources, including wearables, medical equipment, and patient records. This data is processed locally using edge computing devices, eliminating the need for centralized processing.

2. **Data analysis:** The second stage entails analyzing the gathered data using artificial intelligence and machine learning algorithms to find patterns and produce insights. By using this information, healthcare professionals can spot potential health issues before they become serious and provide individualized treatment programs.

3. **Storage and encryption of analyzed data:** The third step entails encrypting and storing the analyzed data on a decentralized blockchain network. This ensures that the data is secure, immutable, and only accessible to authorized users.

4. **Data sharing:** In the fourth stage, healthcare professionals can collaborate and better coordinate patient care by sharing encrypted data across numerous

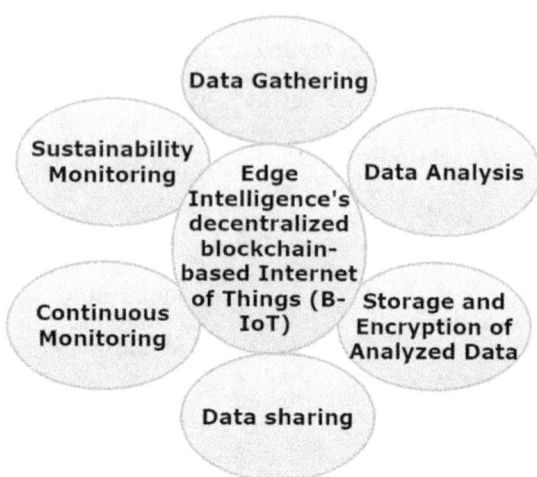

**FIGURE 7.3**   Proposed methodology.

platforms and devices. This could improve patient outcomes while simultaneously reducing healthcare costs.

5. **Continuous monitoring:** The fifth step uses remote patient monitoring
   equipment to assess the patient's condition continuously. As a result, the
   requirement for hospitalization is reduced, and patient outcomes are improved.
   This enables medical practitioners to respond swiftly if a patient's health
   status changes.
6. **Sustainability monitoring:** The last step entails keeping track of how
   healthcare delivery affects the environment and looking for ways to save
   money and energy. The environmental impact of healthcare can be lessened
   and sustainable healthcare practices can be encouraged.

Security, privacy, and sustainability are given priority in the decentralized technique for B-IoT for sustainable healthcare. By implementing this strategy, healthcare
professionals can enhance patient outcomes, lower healthcare expenses, and foster
social and environmental responsibility.

The following steps will be taken to implement a proposed methodology for creating smart contract of Edge Intelligence decentralized B-IoT stakeholders to provide
sustainable healthcare.

**Smart Contract:** Participant Management Contract of B-IoT stakeholders to provide sustainable healthcare

**Data:** The map of the address and corresponding reputation of B-IoT stakeholders
to provide sustainable healthcare

Step 1: Registration of B-IoT stakeholders to provide sustainable healthcare
Step 2: The devices upload their addresses with their unique ID's List (UL)

Step 3: Add unique IDs and the default corresponding reputation value R=0 if they are not registered before

Step 4: Claim Reputation Status (RS)

Step 5: The unique IDs are passed by the stakeholder (SPass)

Step 6: Get the RS values for each stakeholder in UL

Step 7: Return the RS

Step 8: Update RS

Step 9: The UL, the corresponding stakeholder's contributions passed by SPass

Step 10: Update the RS values in UL for Unique ID according to its contribution.

## 7.6  ADVANTAGES

Edge Intelligence decentralized blockchain-based Internet of Things (B-IoT) for sustainable healthcare has several benefits. Here are a few examples:

1. **Improved patient outcomes:** Patient outcomes are improved because of B-IoT technology, which enables healthcare professionals to gather, analyze, and share patient data instantly. This can aid in the early detection of potential health issues, creating individualized treatment strategies, and improving patient outcomes [11].

2. **Cost reduction:** By implementing B-IoT technology, healthcare providers can lower the price tag on emergency department visits, hospital stays, and pointless medical treatments. Early intervention can be made possible via B-IoT, lowering the need for more expensive treatments.

3. **Security and privacy:** B-IoT uses decentralized blockchain technology to ensure patient data is safe and impenetrable. With this technology, patients may take more control over their medical data while maintaining privacy and confidentiality [12].

4. **Better coordination:** B-IoT enables healthcare workers to collaborate and successfully coordinate care, which saves effort duplication and enhances patient outcomes. This may lead to better medical results and cheaper medical costs.

5. **Sustainability:** B-IoT promotes sustainable healthcare practices by enabling medical professionals to monitor the environmental effects of treatment and identify waste reduction and energy efficiency opportunities. This could decrease the damaging effects of healthcare on the environment and promote civic engagement.

6. **Remote patient monitoring:** B-IoT enables remote patient monitoring, which can assist patients in better managing their health and reducing the need for hospitalization. This could include better patient outcomes and decreased healthcare costs [13,14,15].

## 7.7  CHALLENGES

The successful application of Edge Intelligence decentralized blockchain-based Internet of Things (B-IoT) for sustainable healthcare presents many obstacles. A few of them are presented here.

1. **Interoperability:** In order to facilitate seamless data sharing and communication between various systems, B-IoT devices and platforms must be interoperable. Employing several standards and protocols by various systems can make it difficult to achieve interoperability [11].
2. **Data security and privacy:** Sensitive patient data is collected, analyzed, and shared as part of the B-IoT. To increase user confidence in the system, it is crucial to guarantee the confidentiality and privacy of this data. Although it presents issues with scalability and performance, blockchain technology can assist in data security and privacy [12].
3. **Regulation:** The usage of B-IoT for sustainable healthcare must adhere several regulatory standards because the healthcare sector is highly regulated. It can be difficult to establish a uniform strategy across the board because these criteria can differ from one place to another.
4. **Technical expertise:** The IoT uses cutting-edge technologies like edge computing, blockchain, and artificial intelligence. Healthcare providers may lack the technical know-how necessary to adopt and use these technologies [13] effectively.
5. **Cost:** Investing in hardware, software, and training are just a few of the considerable up-front expenditures associated with implementing B-IoT for sustainable healthcare. This may challenge healthcare professionals, particularly in areas with low resources.
6. **Ethical issues:** The application of B-IoT in sustainable healthcare creates issues concerning data ownership, patient permission, and the application of artificial intelligence algorithms. To address these ethical issues, healthcare professionals need to adopt policies and procedures.

## 7.8  OPEN SOURCE TOOLS

Edge Intelligence decentralized blockchain-based Internet of Things (B-IoT) for sustainable healthcare is supported by a number of open-source solutions. Here are a few examples.

1. **Apache NiFi:** Apache NiFi, an open-source data integration solution, provides a web-based interface for organizing and managing data flows. It can be used to collect, process, and transfer data in B-IoT applications.
2. **Eclipse IoT:** Eclipse IoT is an open-source community that provides various tools and frameworks for developing B-IoT applications. It consists of programs such as Eclipse Mosquitto, Eclipse Kura, and Eclipse Hono that provide messaging, device management, and data processing tools.

3. **Docker:** Docker is an open-source framework for building, deploying, and managing distributed applications. We provide a container-based deployment solution for B-IoT applications that simplifies app deployment and maintenance in various environments.

4. **Hyperledger Texture:** The open-source blockchain innovation known as Hyperledger Texture offers measured and adaptable engineering for building B-IoT applications. Its plan upholds keen contracts and conveyed apps.

5. **TensorFlow:** TensorFlow is an open-source program library for manufactured insights and machine learning. It offers a collection of instruments and Application programming interfaces (APIs) to make and refine machine learning models connected to B-IoT applications.

6. **OpenMRS:** OpenMRS is an open-source electronic therapeutic record framework that serves as an establishment for overseeing quiet information. It can be utilized in B-IoT applications to handle and ensure a securely understanding of information.

## 7.9  CASE STUDY

Case studies of Edge Intelligence decentralized blockchain-based Internet of Things (B-IoT) are being used for sustainable healthcare in the following countries:

1. **United States:** Philips and Emory Healthcare collaborated to create a farther seriously care unit (ICU) observing framework utilizing B-IoT innovation. The system's real-time quiet checking within the ICU permits for early discovery of well-being issues and preventive measures. The arrangement utilizes Philips' eICU stage, a telehealth device that empowers inaccessible quiet checking in serious care units. The stage is connected to Emory Healthcare's electronic well-being record (EHR) framework, giving healthcare specialists real-time access to quiet information. Information from persistent screens, restorative gadgets, and other sources are accumulated by the eICU stage utilizing B-IoT innovation. The data is safely sent to the eICU center, where prescient analytics calculations are utilized to distinguish patients who are likely to encounter well-being issues. The approach has appeared to significantly taken a toll on reserve funds and advancements in quiet results. ICU mortality was diminished by 27%, and the length of remain was diminished by 20%, agreeing to a study that was distributed in the Diary of the American Restorative Affiliation. Also, the method diminished healthcare costs by $4,000 per quiet. The potential of Edge Intelligence's decentralized, blockchain-based Internet of Things (B-IoT) for maintainable healthcare within the USA is outlined by this case ponder. By using B-IoT advances, healthcare suppliers may provide superior care, upgrade understanding results, cut costs, and progress supportability.

2. **United Kingdom:** With B-IoT technology, Medicalchain has created a blockchain-based telemedicine platform offering secure and open patient data access. The technology enables users to securely and carefully restrict access to healthcare providers while allowing them to preserve their health

records on a blockchain. Using B-IoT technology, the platform gathers data from many sources, including wearables, medical devices, and mobile apps. The information is safely transferred to the Blockchain, where it is preserved and, with the patient's consent, can be viewed by healthcare professionals. The platform also uses smart contracts to provide patients access control over their medical information. Patients can set access levels for various healthcare professionals, including physicians, nurses, and chemists. The platform has shown considerable reductions in healthcare expenses and improvements in patient outcomes. Medicalchain claims that the platform provides up to a 50% reduction in patient wait times and a 90% reduction in healthcare costs. The potential of Edge Intelligence's decentralized, blockchain-based Internet of Things (B-IoT) for sustainable healthcare in the UK is illustrated by this case study. Healthcare providers may deliver better care, enhance patient outcomes, lower healthcare costs, and support sustainability by utilizing Blockchain and B-IoT technology.

3. **Australia:** In affiliation with UNSW and IBM, a B-IoT stage for checking and assessing water quality within the Extraordinary Boundary Reef has been created. The platform uses B-IoT sensors to gather information on water quality, counting temperature, saltiness, and broken-up oxygen levels. The information is examined utilizing prescient analytics calculations after being safely exchanged to the IBM cloud. The models can recognize potential dangers to the well-being of the reef, such as contamination or coral fading. The stage may offer assistance to moderate the Incredible Boundary Reef, a pivotal environment that underpins various marine species and offers critical financial benefits for the locale. This case ponder illustrates the potential of Edge Intelligence's decentralized, blockchain-based Web of Things (B-IoT) for feasible healthcare in Australia. Using B-IoT innovations, healthcare suppliers can grow their reach in conventional healthcare settings and address broader natural and open well-being concerns. In this case, the B-IoT stage can advance maintainability and offer assistance to ensure a pivotal environment.

4. **South America:** Mediledger has created a blockchain-based stage for following the pharmaceutical supply chain. The stage utilizes B-IoT innovation to gather information on temperature and stickiness levels amid pharmaceutical capacity and travel. With the help of innovation, the Colombian government can screen each step of the supply chain, from generation to conveyance, to guarantee that drugs are transported and put away legitimately and do not ruin or terminate. Moreover, the innovation makes a difference in halting extortion and falsifying. By boosting get to secure and compelling pharmaceuticals, the stage can potentially upgrade open well-being in numerous locales of South America. By disposing of extortion and decreasing squandering, the stage can make strides in understanding results, limit healthcare costs, and develop maintainability. This case illustrates the potential of Edge Intelligence's decentralized blockchain-based Internet of Things (B-IoT) for long-term healthcare in South America. By using blockchain and B-IoT advances, healthcare suppliers may address noteworthy open

well-being issues, such as the accessibility of secure and viable solutions, and advance maintainability by decreasing squandering and killing extortion.

5.  **Russia:** Insilico Medicine has created a B-IoT stage for checking and evaluating Moscow's discussed quality. Utilizing B-IoT sensors, the stage collects information on discussed quality, counting levels of carbon monoxide and particulate matter. Insilico Medicine's cloud stage gets the information securely, which machine learning calculations at that point examine. The calculations can distinguish potential well-being dangers, such as cardiovascular or respiratory issues associated with destitute discussed quality. The stage may make strides in opening well-being in Moscow, which has a few of the most exceedingly bad discussed contamination worldwide. By checking discussed quality and distinguishing potential well-being risks, the stage can assist the Moscow government to focus on activities to make strides in discussing quality and diminish the hazard of related well-being conditions. This case ponder illustrates the potential of Edge Intelligence's decentralized, blockchain-based Internet of Things (B-IoT) for long-term healthcare in Russia. By using B-IoT technology, which empowers them to function exterior of conventional clinic settings, healthcare professionals can address more critical natural and open well-being concerns. In this case, the B-IoT stage can offer assistance move forward open well-being results by tending to discuss contamination and advancing maintainability by decreasing the hazard of related well-being issues.

6.  **India:** MyCrop Innovations has created a B-IoT stage for accuracy in horticulture that employs B-IoT sensors to gather information on soil quality, temperature, stickiness, and other components that influence development. Based on the information, ranchers can use innovation to decide when to plant and gather crops as well as how much water and compost to utilize. By upgrading trim administration, the stage can increment trim yields, diminish squandering, and offer assistance to India in addressing its issues with nourishment security. The arrangement records the supply chain for agrarian items from generation to dissemination utilizing blockchain innovation. This makes it conceivable for the Indian government to keep an eye on the effectiveness and security of rural items as well as to spot and halt extortion and falsification. This case serves as an illustration of the potential of the decentralized, blockchain-based Web of Things (B-IoT) from Edge Insights for Indian healthcare maintainability. Healthcare suppliers may address imperative open well-being issues like nourishment security by utilizing blockchain and B-IoT innovations. This innovation, moreover, advances maintainability by diminishing squandering and extortion.

7.  **China:** In order to accumulate data about patients' well-being and ways of life, counting their levels of physical action, resting propensities, and dietary admissions, CareVoice has created a B-IoT stage for healthcare protection. This stage gives well-being safeguards and the capacity to modify protection plans concurring with a person's well-being and way of life choices and advance solid propensities. By advancing solid practices, the stage can

spare healthcare costs and progress open well-being results in China. The stage employs blockchain innovation to defend the understanding of data and encourage data sharing between protections and healthcare suppliers. Empowering them to form data-driven choices with respect to arrangement plans and estimating makes a difference in safeguards to offer their clients healthcare administrations that are more successful and proficient. This case ponder illustrates the potential of Edge Intelligence's decentralized blockchain-based Web of Things (B-IoT) in China's Maintainable Healthcare. By leveraging blockchain and B-IoT innovation, healthcare specialists can advance solid propensities, decrease healthcare costs, and improve open well-being results.

8.  **South Korea:** The "Walk" may be a B-IoT wearable contraption made by Zikto that utilizes B-IoT sensors to track the walk and balance of more seasoned individuals in real time. In the case of a drop or other mischance, the gadget can recognize irregular walking designs and inform carers or the crisis workforce. Blockchain innovation is additionally utilized by the Walk contraption to securely store and transmit understanding information to carers, family individuals, and healthcare experts. As a result, healthcare suppliers can screen patients remotely and give individualized care whereas keeping up the secrecy and security of quiet data. The Walk contraption can help bring down healthcare uses and improve open well-being results in South Korea by empowering sound maturing and diminishing falls and mischances. The innovation advances maintainability by permitting inaccessible understanding checking and lessening the requirement for in-person restorative arrangements. The potential of Edge Intelligence's decentralized, blockchain-based Web of Things (B-IoT) for South Korea's maintainable healthcare is outlined in this case. By advancing solid maturing, lessening healthcare costs, and empowering inaccessible quiet checking, healthcare suppliers can utilize blockchain and B-IoT innovation to progress open well-being results and advance maintainability.

9.  **Japan:** The organization between the Japanese government and the company Acroquest Innovation is one occasion of how Edge Intelligence's decentralized blockchain-based Web of Things (B-IoT) for economical healthcare is being utilized in Japan. Acroquest Technology's B-IoT stage employs B-IoT sensors to assemble data on persistent conduct and well-being state, counting blood weight, heart rate, and levels of physical movement. With the utilization of the computer program, therapeutic experts may remotely screen patients and spot well-being issues sometime recently, but they decline. This will improve understanding of results and quality of life while lessening clinic confirmations and readmissions. The stage makes utilization of blockchain innovation to safely store and share therapeutic information between healthcare suppliers and patients, securing the security and security of persistent data. The Acroquest Innovation stage can help Japan bring down healthcare consumption and improve open well-being results by empowering inaccessible persistent checking and early conclusion of well-being concerns. The innovation

**TABLE 7.1**
**Comparative Analysis of Different Security Techniques in Edge Intelligence Decentralized B-IoT**

| References | Edge Intelligence | Decentralized | Sustainable Healthcare | Blockchain-Based Internet of Things (B-IoT) |
|---|---|---|---|---|
| [11] | X | ✓ | X | ✓ |
| [12] | X | ✓ | X | ✓ |
| [13] | X | ✓ | X | ✓ |
| [14] | X | ✓ | X | ✓ |
| [15] | X | X | X | ✓ |
| Proposed | ✓ | ✓ | ✓ | ✓ |

bolsters supportability by empowering further understanding checking and limiting the requirement for in-person specialist visits. This case illustrates the potential of Edge Intelligence's decentralized, blockchain-based Web of Things (B-IoT) for long-term healthcare in Japan. Healthcare experts can improve open well-being results and progress supportability by advancing inaccessible persistent checking, diminishing healthcare costs, and encouraging early infection conclusion.

Table 7.1 presents a comparative analysis of different security techniques in Edge Intelligence decentralized blockchain-based Internet of Things (B-IoT).

## 7.10  IMPLEMENTATION

The proposed methodology is implemented using Eclipse Kura, which is an Extensible open source IoT Edge Framework and gateway (Figure 7.4). Docker is installed before using Eclipse Kura (Figure 7.5). The hardware and software components, that is, x86-based Docker machine, Node MCU, docker, Portaner, Kapua, Kura, Mosquitto, Ethereum for blockchain implementation, and Streamsheets, were used.

## 7.11  FUTURE PERSPECTIVE

Edge Intelligence's (Edge) decentralized blockchain-based Internet of Things (B-IoT) for sustainable healthcare has a bright future. Here are some prospective perspectives for the future:

1.  **Better patient results:** Real-time patient monitoring with B-IoT can help identify health risks early and take preventative action. This may result in better patient outcomes and lower medical expenses [11].

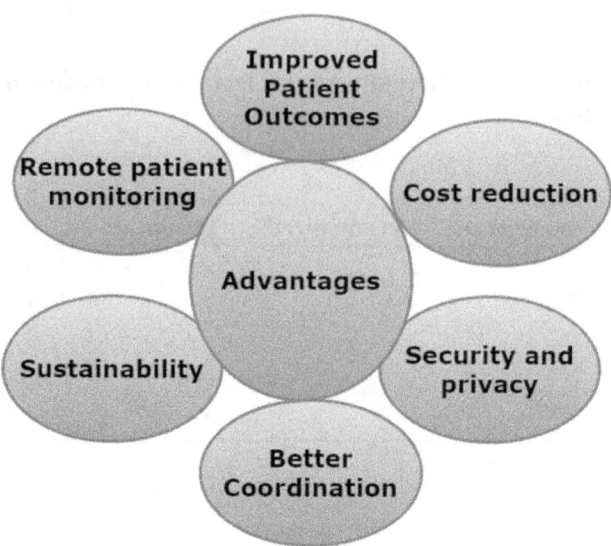

**FIGURE 7.4**  Implementation of the proposed methodology using Eclipse Kura.

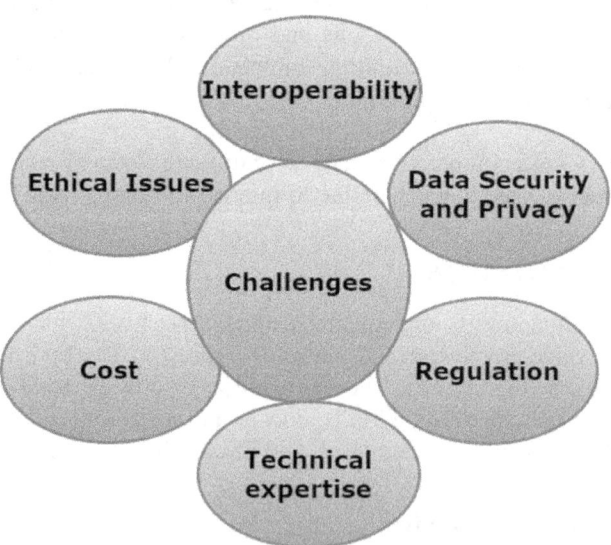

**FIGURE 7.5**  Implementation of the proposed methodology using Wiregraph.

2. **Personalized medicine:** The development of personalized medicine methods can be made possible by the collection and analysis of vast volumes of patient data made possible by the IoT. Healthcare professionals can enhance outcomes and lessen negative effects by customizing therapies for each patient [12].

3. **Predictive analytics:** The construction of predictive analytics models is made possible by the ability of B-IoT to enable the collection of data from a variety of sources. Healthcare professionals can forecast disease outbreaks, identify high-risk patients, and create tailored interventions by looking for patterns in data.

4. **Technological advancement:** The B-IoT is still in its infancy but has the potential to advance significantly. As technology develops, B-IoT might become more scalable, secure, and economical, increasing the likelihood that healthcare providers will use it.

5. **Optimal resource utilization:** B-IoT can assist healthcare providers in making the most efficient use of their resources, reducing waste and boosting efficacy. For instance, by monitoring the use of medical equipment, healthcare professionals can uncover ways to save costs and improve patient care.

6. **Global health:** B-IoT can assist medical practitioners in reaching out to disadvantaged communities in far-flung areas, improving care access, and closing treatment gaps. By utilizing B-IoT technologies, healthcare providers can give real-time health monitoring and remote access to medical experts [13].

## 7.12   CONCLUSION

In this chapter, it is discussed that Edge Intelligence's decentralized blockchain-based Internet of Things (B-IoT) has the potential to revolutionize sustainable healthcare by delivering real-time monitoring, personalized medicine, predictive analytics, effective resource utilization, global health outreach, and technological advancements. B-IoT can help medical professionals deliver better treatment, enhance patient outcomes, cut costs, and advance sustainability. Implementing B-IoT, however, is not without its difficulties, including legislative obstacles, interoperability problems, and privacy and security risks with data. Nevertheless, the B-IoT has a bright future in sustainable healthcare due to open-source tools and ongoing technological breakthroughs. Healthcare providers may advance sustainability and improve healthcare accessibility, effectiveness, and efficiency by utilizing the B-IoT.

## REFERENCES

1.   Dai, Hong-Ning, Zibin Zheng, and Yan Zhang. "Blockchain for internet of things: A survey." *IEEE Internet of Things Journal* 6, no. 5 (2019): 8076–8094.
2.   Dwivedi, Sanjeev Kumar, Priyadarshini Roy, Chinky Karda, Shalini Agrawal, and Ruhul Amin. "Blockchain-based internet of things and industrial IoT: A comprehensive survey." *Security and Communication Networks* 2021 (2021): 1–21.
3.   Dey, Tushar, Shaurya Jaiswal, Shweta Sunderkrishnan, and Neha Katre. "HealthSense: A medical use case of Internet of Things and blockchain." In *2017 International Conference On Intelligent Sustainable Systems (ICISS)*, pp. 486–491. IEEE, 2017.
4.   Bhuiyan, Mohammad Nuruzzaman, Md Mahbubur Rahman, Md Masum Billah, and Dipanita Saha. "Internet of Things (IoT): A review of its enabling technologies in healthcare applications, standards protocols, security, and market opportunities." *IEEE Internet of Things Journal* 8, no. 13 (2021): 10474–10498.

5. Albreem, Mahmoud A., Abdul Manan Sheikh, Mohammed H. Alsharif, Muzammil Jusoh, and Mohd Najib Mohd Yasin. "Green internet of things (GIoT): Applications, practices, awareness, and challenges." *IEEE Access* 9 (2021): 38833–38858.

6. Liu, Yinqiu, Kun Wang, Kai Qian, Miao Du, and Song Guo. "Tornado: Enabling blockchain in heterogeneous internet of things through a space-structured approach." *IEEE Internet of Things Journal* 7, no. 2 (2019): 1273–1286.

7. Restuccia, Francesco, Salvatore D. Kanhere, Tommaso Melodia, and Sajal K. Das. "Blockchain for the Internet of Things: Present and future." arXiv preprint arXiv:1903.07448 (2019).

8. Aljumah, Abdullah, and Tariq Ahamed Ahanger. "Blockchain-Based information sharing security for the Internet of Things." *Mathematics* 11, no. 9 (2023): 2157.

9. Wang, Weiyi, Jin Chen, Yutao Jiao, Jiawen Kang, Wenting Dai, and Yuhua Xu. "Connectivity-aware contract for incentivizing IoT devices in complex wireless blockchain." *IEEE Internet of Things Journal* 10 (2023): 10413.

10. Pathak, Aditya, Irfan Al-Anbagi, and Howard J. Hamilton. "TABI: Trust-based ABAC mechanism for Edge-IoT using blockchain technology." *IEEE Access* 11 (2023): 36379.

11. Attkan, Ankit, Virender Ranga, and Priyanka Ahlawat. "A Rubik's cube cryptosystem based authentication and session key generation model driven in blockchain environment for IoT security." *ACM Transactions on Internet of Things* 4 (2023): 1.

12. Bhadula, Shuchi, Sachin Sharma, and Amar Johri. "Hybrid blockchain and IPFS for secure industry 4.0 framework of IoT-based skin monitoring system." In *2023 9th International Conference on Advanced Computing and Communication Systems (ICACCS)*, vol. 1, pp. 41–47. IEEE, 2023.

13. Nautiyal, Neeraj, Piyush Agarwal, and Sachin Sharma. "Rechain: A secured blockchain-based digital medical health record management system." In *2023 4th International Conference on Innovative Trends in Information Technology (ICITIIT)*, pp. 1–6. IEEE, 2023.

14. Sharma, Ishita and Sachin Sharma. "Blockchain enabled biometric security in Internet-of-Medical-Things (IoMT) devices." In *2022 International Conference on Augmented Intelligence and Sustainable Systems (ICAISS)*, pp. 971–979. IEEE, 2022.

15. Tabaghchi, Milan Sara, Mehdi Darbandi, Nima Jafari Navimipour, and Senay Yalcın. "An energy-aware load balancing method for IoT-based smart recycling machines using an artificial chemical reaction optimization algorithm." *Algorithms* 16, no. 2 (2023): 115.

# 8 Analysis of AI Embedded Block Chain Security Model for Healthcare and Financial Transactions

*Jeya Mala[1] and A. Pradeep Reynold[2]*

[1] School of Computer Science and Engineering, Vellore Institute of Technology, Chennai, India

[2] Department of Safety, ASET College of Science and Technology, Chennai, India

## 8.1 INTRODUCTION

Errors and fraud in the healthcare system ruin everything. The safeguarding of patient data and medical solutions is thus one of the most crucial application of artificial intelligence (AI) in healthcare. In the past, fraud and security breaches could only be uncovered through manual rule execution and system reviews (Abu-elezz et al., 2020). Estimates suggest that $17 billion a year can be saved by increasing the efficiency of fraud detection due to the advent of AI's role in the identification of such breaches. A recent study found that healthcare businesses might lose an average of $380 per patient record due to cybercrime. Trust and loyalty built via AI in healthcare security monitoring and anomaly detection can pave the way for more digital disruption in the healthcare industry. Healthcare records for 5.57 million patients were compromised in at least 477 data breaches in 2017, according to reports of the Department of Health and Human Services (DHHS) or the media. It is estimated that the average cost of a data breach is $408 per compromised record, which is more than 2.75 times the global average. Due in part to this persistent problem, pioneers in healthcare information technology (IT) are investigating how Blockchain may be used to create a safer, more unified healthcare records database (Haleem et al., 2021)

Blockchain technology's decentralized nature and in-built safeguards make it a promising tool for enhancing cybersecurity. To name a few, IBM, MIT, and Walmart are among the many major corporations and academic organizations already seeking to incorporate Blockchain technology into healthcare recordkeeping. Blockchain's digital design, a decentralized chain of blocks, has inherent security advantages. Thousands of patients' records may be at risk because they are all stored in the same digital repository. Even if a hacker could gain the encryption key to one patient's record in a blockchain of patient records, the damage would be mitigated because the hacker would also require each member's individual encryption key. This would still require the hacker to perform the same steps for each patient. This aids in the prevention of widespread, multi-patient breaches (Tariq et al., 2020). Every participant in a Blockchain network can view and verify the accuracy of the full distributed ledger. Thus, anyone with a vested interest can verify and keep their data in individual blocks up to date. Legislation and rules can be included in the Blockchain as smart contracts, which is a major advantage. In the Blockchain, logic rules are stored as "smart contracts." The preceding agreement is automatically binding on all parties in these contracts. Like traditional contracts and regulations, smart contracts can be used to formalize responsibilities and consequences. This allows the Blockchain to be designed to meet the standards of data privacy and security laws such as HIPAA (1996's Health Insurance Portability and Accountability Act) and the EU's General Data Protection Regulation (2016's General Data Protection Regulation) (Hylock and Xiaoming, 2019). Because of this, this chapter examines how Blockchain can be used to secure healthcare data and the areas in which researchers are likely to go next.

## 8.2 BLOCKCHAIN: A BRIEF OVERVIEW

A Blockchain is a decentralized digital ledger that stores transaction data in blocks corresponding to specific timestamps. This ledger is ideal for transparent data storing because of its immutability and limited access.

A shared, decentralized ledger; immutable logs; autonomous means of reaching an agreement; electronic contracts (also known as "smart contracts"); a pair of keys used in cryptography; access control; improved security measures; and a network in which users share resources directly with one anotherare all principles of Blockchain technology. There is no need for a central authority or a trustworthy third party to interfere because all monetary transactions are public and verifiable. Since the system is dispersed and decentralized, all healthcare providers, facilities, and other caregivers would have access to the same unified patient record. This is just one way the Blockchain can lead to quicker diagnoses and more tailored treatment plans. Blockchain's distributed ledger technology could significantly enhance the management of the pharmaceutical supply chain. Once a drug ledger has been constructed, it can track the drug back to its source (i.e., a laboratory). The ledger will then keep track of the information from the point it was first touched, through all of the hands it has passed through before finally reaching the end user. The system can also record consumed materials and waste products.

## 8.3 APPLICATION OF BLOCKCHAIN IN CYBERSECURITY

Cybersecurity is the technique of keeping computers and networks safe against hackers who may try to gain unauthorized access to modify or destroy data to blackmail the owner or get access to private information. As more and more of our lives become dependent on digital data and transactions, the significance of taking extra precautions to keep these areas safe grows. Computer system attacks frequently use malicious software like viruses, Trojan horses, rootkits, etc. Common cyberattack methods include phishing, man-in-the-middle attacks, distributed denial-of-service attacks, Structured Query Language (SQL) injection, and ransomware.

Features crucial to cybersecurity are the following:

1. Checksums in cryptography
2. Codes for restoring and correcting data
3. Take into account potential dangers
4. Take safety measures to reduce system weaknesses
5. Understanding the characteristics of malicious code
6. Controlling who has entry is item number six in the realm of access management
7. Authentication
8. Encryption
9. Establishing barriers
10. Prevention of intrusions using intrusion detection systemsand intrusion prevention systems.

## 8.4   BLOCKCHAIN-BASED CYBERSECURITY APPLICATIONS

### 8.4.1   SAFEGUARDING IoT DEVICES

Data and systems security from cyber criminals has always been a major worry as AI and IoT applications grow. For example, Blockchain might be put to work to keep the IoT system secure by encrypting communications between devices, safeguarding keys, and authenticating users (Xiaohua et al., 2021).

### 8.4.2   RELIABILITY OF DIGITALLY DISTRIBUTED PROGRAMS

By verifying updates and installers with Blockchain technology, we can lower the likelihood that malicious software will exploit compromised PCs. Here, hashes are recorded in the Blockchain so that new software identities may be checked against them to ensure they are legitimate.

### 8.4.3   TRANSMISSION SECURITY

You may protect your data from prying eyes by encrypting it before sending it over a network.

Essential data is stored in a decentralized manner.

Given the exponential growth in data creation, Blockchain-based storage solutions are a promising way to ensure digital information security while accommodating the vast quantities of data that will inevitably be generated (Clim et al, 2019).

### 8.4.4 Preventing DoS Attacks

One of the most popular forms of cyberattack today is the distributed denial of service (DDoS) attack. Hackers launch DoS attacks to clog up the Internet and slow down service provisioning. Because of its immutability and cryptographic properties, Blockchain has the potential to be an effective defense mechanism against these dangers. Comparable to a public directory, the Domain Name System (DNS) associates IP addresses with domain names. Over time, hackers have tried to get into the DNS and use these vulnerabilities to redirect traffic to their own sites. Due to Blockchain's immutability and decentralized nature, the DNS can be stored with greater security. Secrecy means only those who have permission to view the information are granted access to it. Blockchain data can be fully encrypted to shield it from snoops even as it flows through less-than-secure channels. To prevent attacks from within the network, it is required to implement security measures at the application level, such as access controls. Blockchain technology can provide increased security due to the usage of public key infrastructure for identity verification and communication encryption. But if duplicate private keys are kept in a different location, there's a significant chance that they'll be stolen. Cryptographic methods based on integer factorization and other key management techniques are advised as a safeguard against this. Blockchain's immutability and auditability are two properties that help firms ensure the reliability of their data. And in the event of a 51% cyber control attack, businesses can benefit from consensus model protocols that help implement safeguards against and management of ledger splitting. Since the system's previous state is recorded with each new iteration in Blockchain, the ledger can be thoroughly verified at any time. Smart contracts allow for the enforcement of agreements between parties to ensure that miners cannot steal information from already-created blocks.

It's safe to say that DDoS attacks, which aim to make technological services unavailable to users, are the most common type of cyberattack in use today. Distributed ledger technology (Blockchain) mitigates this problem by requiring the attacker to do a huge number of very small transactions to overwhelm the network, which is inefficient and time-consuming. The lack of a centralized processing unit in Blockchain reduces the possibility that an IP-based DDoS attack will cause interruption. Due to data persistence across multiple nodes, complete copies of the ledger are always available. The distributed architecture and utilization of several nodes contribute to the stability of these systems and platforms.

## 8.5 INFORMATION ASSURANCE IN A CYBER PANDEMIC USING BLOCKCHAIN

Keeping in touch with loved ones during the epidemic was challenging, but due to modern technology, we could do so, providing an opportunity for human connection that post-apocalyptic generations lacked. This increased connectivity, however, opens new doors for the exploitation of vulnerabilities that have always been present but were previously unknown to the public.

Cybercrime has increased by 600% over the previous year, according to the 2021 Cyber Security Trends Report published by PurpleSec, a company that offers both offensive and defensive cybersecurity services. Consumers' unfamiliarity with phishing has been blamed for the uptick; hackers employed WHO and CDC-looking email addresses to trick people into visiting malicious websites or downloading malware. However, the information on our computers isn't the only thing at risk from these attacks. According to the US Department of Health and Human Services, healthcare data breaches affected over 1 million people every single month of 2020 (Landi, 2022). Since 2019, this has contributed to a rise of 9,851% in the number of healthcare data breaches, affecting over 630 different businesses. These violations hampered healthcare providers during the first pandemic of the new millennium, resulting in the deaths of millions of individuals who may have been rescued. Since we did so well with the corona virus, it is imperative that we apply the same amount of pressure and aid to the pharmaceutical industry and research facilities in order to find a solution to cybersecurity challenges. Blockchain, thankfully, is a new and encouraging cure for the problems of technology. Blockchain, as defined by the National Institute of Standards and Technology, is a "tamper evident and tamper resistant digital ledger" (or DLT) that allows distributed ledgers of data to be maintained and shared between systems without the need for a central authority or repository. Blockchain's use in this context eliminates "many dangers connected with centrally stored data," as Infosys associate consultant Yogesh Shelke pointed out. This is especially important given the prevalence of attacks on centralized medical databases. Even if an attacker gains access to the medical data stored in one of these DLTs, they would have a hard time, computationally speaking, altering it or holding it for ransom. Blockchain does this by utilizing a unique design in which information is stored in "blocks" and cryptographically "chained" together. This means that any attempt to alter the contents of the block would result in a new cryptographic hash, breaking the block's chain and rendering any subsequent updates to the medical data meaningless. Similar to how there are copies of the DLT on many nodes (computers connected to the Blockchain network), this ensures that hospital managers always have access to their patient's data safely and securely, regardless of their physical location. In addition, Blockchain may help patients regain control over their own data and the therapeutic decisions that are made because of that data. With the use of smart contracts, patients may control the distribution of their medical data in a way that best suits their needs, be it through a more streamlined, efficient process or a more nimble, self-directed effort (programs that automate the implementation of agreements between parties without an intermediary). This type of secure data transfer, in conjunction with multi-party computation technology, which safeguards the privacy of the parties involved in the smart contract's transaction, can be used to piece together the shards of patient data stored in various record systems to form a complete picture of the patient's health.

## 8.6 REAL-WORLD ILLUSTRATIONS OF BLOCKCHAIN IN CYBERSECURITY

The following are examples of some of the most significant practical uses of Blockchain in cybersecurity today (Mathew, 2019).

Conventional bank Barclays (London, England) has applied for a patent on a method of using Blockchain technology to increase the security of financial transactions. It makes use of distributed ledger technology (DLT) to stabilize Bitcoin exchanges (DLT). Consequently, the bank can store customer information securely on the Blockchain.

1. Cisco (San Jose, CA) is considering utilizing Blockchain to safeguard IoT devices since it eliminates network bottlenecks and encrypts data.
2. A Bitcoin trading platform situated in San Francisco Coin base takes extra precautions to protect its customers' privacy by encrypting sensitive data like passwords and wallet addresses. Further, it ensures the security of its encryption by doing extensive screenings of its staff.
3. Canberra, Australia's capital, is considering utilizing DLT to build a nationwide system of cyber defense. The government and IBM have worked together to create a Blockchain ecosystem for the secure storage of official documents.
4. Philips Healthcare (Andover, Massachusetts), which has collaborated with healthcare facilities worldwide to create a Blockchain and AI-based healthcare ecosystem. The discovery and analysis tools in this ecosystem will improve the availability of a wide range of operational, administrative, and medical data.
5. The defense and military establishments in Beijing, China are experimenting with Blockchain security to safeguard government and military data and intelligence.
6. With the goal of becoming the world's first decentralized bank, Founders Bank in Valletta, Malta, is now taking bitcoin deposits. Ideas like encryption and distributed ledgers will be used to secure crypto currency storage.
7. The Government of Colorado (Denver, Colorado) is considering deploying Blockchain technology to protect data and fight the growing number of cyberattacks, as stipulated by a law recently adopted by the state's senate.
8. J.P. Morgan's New York office employs Blockchain technology on its own platform Quorum to complete private financial transactions. It utilizes encryption and the concepts underpinning Smart contracts to guarantee the confidentiality of all business conducted.

In Mountain View, California, Health Linkages is developing a Blockchain-based system to restrict access to patient data to authorized personnel only. It will also be used to record significant medical developments throughout time, giving clinicians a complete picture of a patient's health background (Shi et al., 2020).

## 8.7 EMBEDDING AI IN BLOCKCHAIN FOR ENHANCED SECURITY

While AI and Blockchain are very different in nature, they do complement one another and can be used independently or together to greatly improve cybersecurity (Dinh and Thai, 2018). In this age of digitalization, the most recent attack vectors must be mitigated, making cybersecurity paramount. There is no sector of the economy that is immune to cybercrime .

### 8.7.1 Cybersecurity Threat Detection Using Artificial Intelligence and Blockchain

The methods we use to prevent cyberattacks today are lifesaving, but they also give rise to increasingly sophisticated attacks. According to data compiled for the "2018 SonicWall Cyber Threat Report," the number of cyberattacks and malware variants has increased by 101.2% over the previous year. Cybercriminals are upgrading to more sophisticated methods of attacking businesses online. According to Toshiba, 62% of European IT executives rank data security as a top investment priority.

The cost of data breaches worldwide rose by 6.4% in 2018 compared to 2017, according to a study by IBM and the Ponemon Institute. As more and more of a company's operations move online, the price of any security lapse will rise accordingly.

Laptops and their users are often viewed as an organization's first line of defense against unauthorized access to sensitive data; however, this can be compromised in a number of ways, including when devices are misplaced, stolen, or allowed to connect to unsecure networks.

The proliferation of IoT raises new security challenges for businesses, heightening the importance of robust privacy safeguards. Since businesses aren't doing enough to protect themselves online, we may expect to see an uptick in both the severity and frequency of cyberattacks.

### 8.7.2 Use of Artificial Intelligence in Cybersecurity

Virtually all currently available cybersecurity tools appear to use some form of AI. It's possible you'll notice that some specialists talk about a breakthrough while others try to build on the buzz. Despite the fact that this is a widely discussed topic, nearly all security professionals believe that AI is the key to resolving today's cybersecurity problems. But the question now is how significant the possibilities and repercussions of this new technology really are, given that hackers gain from new AI approaches.

### 8.7.3 The Importance of AI in Cybersecurity

It is clear from a review of the security architecture that many businesses have made significant investments in security operations centers (SOCs) and other components of the security infrastructure in order to extract greater value from these areas. Companies still have difficulties protecting their enterprises and data. This is because technological limitations, amplified by the attacks' sophistication, place severe constraints on internal processes and security teams.

Consequently, the technological limits and processes must be addressed so that SOC steams can even stay ahead of the most advanced hackers. AI processes of investigation and reaction can be attained through enhanced performance, accuracy, and management of enormous datasets, but only if disparate tools are integrated into a single security architecture.

In addition, the exponential growth in the variety and quantity of malware and the extent of the Darknet has pushed us to the boundaries of what can be accomplished with signatures and heuristics. Signatures represent the malicious code's fingerprint and aid in detecting and identifying malware. One of the problems with this method is that it often fails to detect changes to malware signatures when just minor elements are altered. Some security researchers claim that current anti-virus signatures only protect against 30–40% of malware, while some put that number as high as 65–70%.

To delete or corrupt a Blockchain, a hacker would need to destroy the data stored on every machine in the network. This might involve millions of computers depending on how much data is being saved. Undamaged computers, often known as "nodes," would continue to check and save all the network data until the hacker brought down the entire network simultaneously, which is highly unlikely. If many people use the network, it will be nearly impossible to shut down the whole chain. There is an immeasurably smaller probability of being hacked into by hackers on larger Blockchain networks with more users. This framework, known as Blockchain technology, allows for the most secure online data storage and exchange possible. Because of this, creative thinkers have started implementing technological solutions to combat fraud and strengthen data security in a variety of industries. The need to double-check keys is something the organization does away with. Instead, every bit of data is dispersed to various nodes in the network. In the event of a data modification attempt, the system examines the entire collection of chains, makes comparisons, and eliminates any that do not correspond to the metadata packet. The primary characteristics of Blockchain are its security and the ability to incorporate AI code onto its platform. Using minor architecture adjustments, this distributed ledger is constructed and developed in order to store millions of records securely on the platform. It is created with its working protocol as proof that each participant in the Blockchain is accountable for all modifications, so that all modifications must be accepted. Similarly, it enables a trustless principle in which all transactions are anonymous, remain on the chain, and identification is stored while maintaining data integrity. Thus, Blockchain represents the next step in the evolution of databases.

Each time a block is solved, a new block is created that contains a hash key of the previous block. This creates an unending chain of blocks. The decentralized nature of Blockchain is a key factor in its security. Blockchain platforms use proof-of-stake-based consensus among all members to validate any changes that can be made to a specific block. This requires users to demonstrate their ownership. Proof-of-stake (PoS) is a method for achieving distributed consensus using a Blockchain network. Proof-of-work (PoW), on the other hand, requires users to repeatedly execute hashing algorithms or other client puzzles in order to validate electronic transactions. Both techniques are considered valid for the use of any Blockchain updates by all users. Hundreds, if not thousands, of users will thereby prevent any external modifications to the Blockchain Intruders.

However, the aspect that has garnered the most attention from cyber security experts is the ability to create smart contracts utilizing a Blockchain platform. In addition to defining the rules and penalties for an agreement, smart contracts automatically enforce all parties' agreements.

Instead of legal language, computer language is used to record the terms of smart contracts. A good distributed ledger system can execute smart contracts automatically. Low contracting, enforcement, and compliance costs are among the probable benefits of smart contracts. Therefore, contracts for numerous transactions of modest value are viable. NuCypher, Acronis, IBM Blockchain platform, and Facebook are companies that raise substantial capital for their Blockchain security platforms. Despite their apparent incompatibility, merging AI and Blockchain technologies could have significant benefits. Now, industries are working to create Blockchain and AI-enabled applications. The two technologies could not be more different from one another; nevertheless, when combined, they will usher in an entirely new paradigm. When AI and Blockchain are utilized together, it offers additional protection against cyber security threats. While AI may be taught to automate the detection of real-time attacks, Blockchain will be used to identify security flaws in decentralized databases.

## 8.8 ASSESSMENT ON THE NEED FOR AI IN BLOCKCHAIN SECURITY FOR HEALTHCARE DATA

For a Hospital scenario, the patients' medicine orders and their payment transactions are done using Blockchain. The case study dataset is taken from Kaggle: www.kaggle.com/datasets/palwashakhanzadi/iot-blockchain-data?select=transaction.csv.

Keeping our essential medical data safe and private is the most popular Blockchain healthcare application, which isn't surprising. Security is a key issue in the healthcare industry. . Six hundred ninety-two significant healthcare data breaches were reported between July 2021 and June 2022. The attackers took credit card and banking information and health, orders, transaction, and account of patients' records.

Blockchain's capacity to preserve an immutable, decentralized, and transparent ledger of all patient data makes it a technology fertile for security applications. Additionally, while Blockchain is visible, it is also private, concealing the identity of any individual with complicated and secure protocols that can preserve the sensitivity of medical data. The decentralized structure of the technology also allows patients, doctors, and healthcare providers to share the same information swiftly and safely (Sharma and Balamurugan, 2020).

This chapter, however, makes an effort to learn how well Blockchain technology functions in its current form and how it might be enhanced through the application of AI and machine learning (ML). The sample data from a voluminous amount of data in the dataset is taken for understanding and is presented in Tables 8.1 and 8.2.

**TABLE 8.1**
**Order Data with Blockchain Id**

| Patient's Device_id | Frequency | Station_to | Account_to | Evidence_id | k_symbol |
|---|---|---|---|---|---|
| 34323 | 1 | 265000 | 582 | 2 | chain1 |
| 76546 | 2 | 263358 | 157 | 1 | chain1 |
| 654567 | 3 | 262000 | 149 | 2.3 | chain1 |
| 7688567 | 3 | 263358 | 582 | 1.83 | chain1 |
| 4566 | 4 | 319000 | 249 | 1.2 | chain1 |
| 34533 | 5 | 263358 | 582 | 0.7 | chain1 |
| 345645 | 5 | 140000 | 553 | 1.7 | chain1 |
| 7687 | 6 | 223000 | 75 | 1.18 | chain1 |
| 68767 | 6 | 263358 | 582 | 1.3 | chain1 |
| 45645 | 6 | 243000 | 5882 | 3.2 | chain1 |
| 3455645 | 7 | 293000 | 69 | 1.3 | chain1 |
| 53435 | 7 | 235000 | 84 | 2.7 | chain1 |
| 64567 | 8 | 254000 | 514 | 1.18 | chain1 |
| 5678 | 8 | 286000 | 53 | 1 | chain1 |
| 56546 | 9 | 252000 | 369 | 1.7 | chain1 |
| 4546 | 9 | 210000 | 737 | 0.7 | chain1 |
| 45634 | 9 | 267000 | 248 | 1.3 | chain1 |
| 34323 | 1 | 265000 | 582 | 2 | chain1 |
| 76546 | 2 | 263358 | 157 | 1 | chain1 |
| 654567 | 3 | 262000 | 149 | 2.3 | chain1 |
| 768 8567 | 3 | 263358 | 582 | 1.83 | chain1 |
| 4566 | 4 | 319000 | 249 | 1.2 | chain1 |
| 34533 | 5 | 263358 | 582 | 0.7 | chain1 |
| 345645 | 5 | 140000 | 553 | 1.7 | chain1 |
| 7687 | 6 | 223000 | 75 | 1.18 | chain1 |
| 68767 | 6 | 263358 | 582 | 1.3 | chain1 |
| 45645 | 6 | 243000 | 5882 | 3.2 | chain1 |
| 3455645 | 7 | 293000 | 69 | 1.3 | chain1 |
| 53435 | 7 | 235000 | 84 | 2.7 | chain1 |
| 64567 | 8 | 254000 | 514 | 1.18 | chain1 |
| 5678 | 8 | 286000 | 53 | 1 | chain1 |
| 56546 | 9 | 252000 | 369 | 1.7 | chain1 |
| 4546 | 9 | 210000 | 737 | 0.7 | chain1 |
| 45634 | 9 | 267000 | 248 | 1.3 | chain1 |
| 34323 | 1 | 265000 | 582 | 2 | chain1 |
| 76546 | 2 | 263358 | 157 | 1 | chain1 |
| 654567 | 3 | 262000 | 149 | 2.3 | chain1 |
| 7688567 | 3 | 263358 | 582 | 1.83 | chain1 |
| 4566 | 4 | 319000 | 249 | 1.2 | chain1 |
| 34533 | 5 | 263358 | 582 | 0.7 | chain1 |
| 345645 | 5 | 140000 | 553 | 1.7 | chain1 |
| 7687 | 6 | 223000 | 75 | 1.18 | chain1 |
| 68767 | 6 | 263358 | 582 | 1.3 | chain1 |

**TABLE 8.1 (Continued)**
**Order Data with Blockchain Id**

| Patient's Device_id | Frequency | Station_to | Account_to | Evidence_id | k_symbol |
|---|---|---|---|---|---|
| 45645 | 6 | 243000 | 5882 | 3.2 | chain1 |
| 3455645 | 7 | 293000 | 69 | 1.3 | chain1 |
| 53435 | 7 | 235000 | 84 | 2.7 | chain1 |
| 64567 | 8 | 254000 | 514 | 1.18 | chain1 |
| 5678 | 8 | 286000 | 53 | 1 | chain1 |
| 56546 | 9 | 252000 | 369 | 1.7 | chain1 |
| 4546 | 9 | 210000 | 737 | 0.7 | chain1 |
| 45634 | 9 | 267000 | 248 | 1.3 | chain1 |
| 34323 | 1 | 265000 | 582 | 2 | chain1 |
| 76546 | 2 | 263358 | 157 | 1 | chain1 |
| 654567 | 3 | 262000 | 149 | 2.3 | chain1 |
| 7688567 | 3 | 263358 | 582 | 1.83 | chain1 |
| 4566 | 4 | 319000 | 249 | 1.2 | chain1 |
| 34533 | 5 | 263358 | 582 | 0.7 | chain1 |
| 345645 | 5 | 140000 | 553 | 1.7 | chain1 |
| 7687 | 6 | 223000 | 75 | 1.18 | chain1 |
| 68767 | 6 | 263358 | 582 | 1.3 | chain1 |
| 45645 | 6 | 243000 | 5882 | 3.2 | chain1 |
| 3455645 | 7 | 293000 | 69 | 1.3 | chain1 |
| 53435 | 7 | 235000 | 84 | 2.7 | chain1 |
| 64567 | 8 | 254000 | 514 | 1.18 | chain1 |
| 5678 | 8 | 286000 | 53 | 1 | chain1 |
| 56546 | 9 | 252000 | 369 | 1.7 | chain1 |
| 4546 | 9 | 210000 | 737 | 0.7 | chain1 |
| 45634 | 9 | 267000 | 248 | 1.3 | chain1 |
| 34323 | 1 | 265000 | 582 | 2 | chain1 |
| 76546 | 2 | 263358 | 157 | 1 | chain1 |
| 654567 | 3 | 262000 | 149 | 2.3 | chain1 |
| 7688567 | 3 | 263358 | 582 | 1.83 | chain1 |
| 4566 | 4 | 319000 | 249 | 1.2 | chain1 |
| 34533 | 5 | 263358 | 582 | 0.7 | chain1 |
| 345645 | 5 | 140000 | 553 | 1.7 | chain1 |
| 7687 | 6 | 223000 | 75 | 1.18 | chain1 |
| 68767 | 6 | 263358 | 582 | 1.3 | chain1 |
| 45645 | 6 | 243000 | 5882 | 3.2 | chain1 |
| 3455645 | 7 | 293000 | 69 | 1.3 | chain1 |
| 53435 | 7 | 235000 | 84 | 2.7 | chain1 |
| 64567 | 8 | 254000 | 514 | 1.18 | chain1 |
| 5678 | 8 | 286000 | 53 | 1 | chain1 |
| 56546 | 9 | 252000 | 369 | 1.7 | chain1 |
| 4546 | 9 | 210000 | 737 | 0.7 | chain1 |
| 45634 | 9 | 267000 | 248 | 1.3 | chain1 |

*(Continued)*

**TABLE 8.1  (Continued)**
**Order Data with Blockchain Id**

| Patient's Device_id | Frequency | Station_to | Account_to | Evidence_id | k_symbol |
|---|---|---|---|---|---|
| 29284 | 1 | 263358 | 7861 | 2.7 | chain2 |
| 765456 | 2 | 388000 | 123 | 1.18 | chain2 |
| 8789 | 3 | 166000 | 582 | 1.1 | chain2 |
| 657 | 4 | 149000 | 148 | 0.8 | chain2 |
| 457576 | 4 | 302000 | 159 | 0.6 | chain2 |
| 34645 | 5 | 153000 | 124 | 1.1 | chain2 |
| 345345 | 6 | 395000 | 129 | 0.7 | chain2 |
| 5756 | 6 | 216000 | 607 | 1.1 | chain2 |
| 6676 | 6 | 119000 | 68 | 1.2 | chain2 |
| 34533 | 6 | 149000 | 224 | 0.9 | chain2 |
| 345645 | 7 | 263358 | 582 | 2.1 | chain2 |
| 34345 | 7 | 181000 | 115 | 0.6 | chain2 |
| 754567 | 8 | 255000 | 59 | 2.9 | chain2 |
| ... | ... | ... | ... | ... | .. |
| ... | ... | ... | ... | ... | ... |

The transaction details along with the transaction id and the corresponding Blockchain id are presented in Table 8.2.

**TABLE 8.2**
**Transaction Table**

| Id | Account_id | Type | Account_to | Evidence | k_symbol |
|---|---|---|---|---|---|
| 75 | 0 | 0 | 582 | 2 | chain1 |
| 65 | 0 | 0 | 157 | 1 | chain1 |
| 87 | 1 | 0 | 149 | 2.3 | chain1 |
| 75 | 0 | 1 | 582 | 1.83 | chain1 |
| 50 | 1 | 1 | 249 | 1.2 | chain1 |
| 70 | 0 | 0 | 582 | 0.7 | chain1 |
| 80 | 1 | 0 | 553 | 1.7 | chain1 |
| 70 | 1 | 8 | 45 | 0.7 | chain1 |
| 45 | 0 | 9 | 546 | 1.8 | chain1 |
| 85 | 0 | 9 | 455656 | 1 | chain1 |
| 70 | 0 | 9 | 345645 | 1.1 | chain1 |
| 44 | 0 | 2 | 65455 | 2.3 | chain1 |
| 63 | 1 | 2 | 64567 | 1.3 | chain1 |
| 60 | 0 | 3 | 45 | 1 | chain1 |
| 50 | 0 | 4 | 546 | 0.7 | chain1 |
| 60 | 1 | 5 | 455656 | 1.8 | chain1 |
| 61 | 0 | 8 | 7564 | 1.2 | chain1 |

**TABLE 8.2  (Continued)**
**Transaction Table**

| Id | Account_id | Type | Account_to | Evidence | k_symbol |
|---|---|---|---|---|---|
| 50 | 1 | 8 | 56756 | 0.9 | chain1 |
| 62 | 0 | 9 | 34536 | 0.8 | chain1 |
| 48 | 1 | 9 | 4556 | 1 | chain1 |
| 70 | 0 | 1 | 23423 | 0.6 | chain1 |
| 69 | 0 | 2 | 75456 | 2 | chain1 |
| 72 | 0 | 3 | 7678 | 1 | chain1 |
| 53 | 1 | 4 | 56756 | 0.9 | chain1 |
| 72 | 0 | 4 | 34536 | 0.9 | chain1 |
| 72 | 1 | 5 | 4556 | 0.8 | chain1 |
| 60 | 1 | 8 | 657 | 0.6 | chain1 |
| 75 | 1 | 9 | 457576 | 1.1 | chain1 |
| 49 | 1 | 9 | 34645 | 0.7 | chain1 |
| 68 | 1 | 1 | 29284 | 2.7 | chain1 |
| 94 | 0 | 2 | 765456 | 1.18 | chain1 |
| 60 | 0 | 3 | 8789 | 1.1 | chain1 |
| 51 | 0 | 8 | 987689 | 0.8 | chain1 |
| 60 | 1 | 9 | 46768 | 1 | chain1 |
| 72 | 0 | 9 | 567867 | 1.1 | chain1 |
| 72 | 1 | 9 | 565567 | 1.1 | chain1 |
| 55 | 0 | 0 | 7861 | 2.7 | chain2 |
| 80 | 1 | 0 | 123 | 1.18 | chain2 |
| 45 | 0 | 0 | 582 | 1.1 | chain2 |
| 80 | 0 | 1 | 148 | 0.8 | chain2 |
| 50 | 1 | 1 | 159 | 0.6 | chain2 |
| 50 | 0 | 1 | 124 | 1.1 | chain2 |
| 57 | 1 | 0 | 129 | 0.7 | chain2 |
| 60 | 1 | 8 | 987689 | 0.8 | chain2 |
| 60 | 0 | 9 | 46768 | 1 | chain2 |
| 65 | 0 | 9 | 45645 | 1.18 | chain2 |
| 67 | 0 | 9 | 3455645 | 1.1 | chain2 |
| 57 | 1 | 2 | 76545 | 1.1 | chain2 |
| 70 | 1 | 3 | 63456 | 1 | chain2 |
| 72 | 1 | 3 | 987689 | 1.2 | chain2 |
| 70 | 1 | 4 | 46768 | 0.8 | chain2 |
| 66 | 1 | 5 | 45645 | 1 | chain2 |
| 53 | 1 | 8 | 45 | 0.7 | chain2 |
| 65 | 1 | 9 | 56546 | 1.7 | chain2 |
| 45 | 1 | 9 | 4546 | 0.7 | chain2 |
| 65 | 1 | 9 | 45634 | 1.3 | chain2 |
| 58 | 1 | 1 | 53435 | 1.1 | chain2 |
| 90 | 1 | 2 | 64567 | 2.7 | chain2 |
| 60 | 1 | 3 | 5678 | 1.18 | chain2 |
| 60 | 0 | 4 | 56546 | 1 | chain2 |
| 45 | 0 | 4 | 4546 | 1.7 | chain2 |

*(Continued)*

**TABLE 8.2 (Continued)**
**Transaction Table**

| Id | Account_id | Type | Account_to | Evidence | k_symbol |
|----|-----------|------|-----------|----------|----------|
| 70 | 0 | 8 | 7678 | 0.9 | chain2 |
| 70 | 0 | 8 | 56756 | 0.9 | chain2 |
| 60 | 1 | 9 | 34536 | 0.8 | chain2 |
| 82 | 1 | 9 | 4556 | 1 | chain2 |
| 53 | 0 | 1 | 23423 | 0.6 | chain2 |
| 85 | 0 | 2 | 75456 | 2 | chain2 |
| 60 | 0 | 3 | 7678 | 1 | chain2 |
| 60 | 0 | 8 | 7688567 | 1.2 | chain2 |
| 95 | 1 | 9 | 4566 | 0.7 | chain2 |
| 45 | 0 | 9 | 34645645 | 0.8 | chain2 |
| 70 | 0 | 9 | 3464565 | 1.2 | chain2 |
| 65 | 0 | 0 | 146 | 0.6 | chain3 |
| 75 | 1 | 0 | 81 | 2 | chain3 |
| 70 | 1 | 0 | 125 | 1 | chain3 |
| 95 | 1 | 0 | 112 | 0.9 | chain3 |
| 65 | 0 | 1 | 94 | 0.9 | chain3 |
| 70 | 0 | 1 | 571 | 0.8 | chain3 |
| 68 | 1 | 0 | 577 | 1 | chain3 |
| 49 | 0 | 8 | 7688567 | 1.2 | chain3 |
| 42 | 1 | 9 | 4566 | 0.7 | chain3 |
| 69 | 0 | 9 | 34533 | 1.7 | chain3 |
| 60 | 1 | 1 | 34323 | 2 | chain3 |
| 70 | 0 | 2 | 76546 | 1 | chain3 |
| 60 | 1 | 3 | 654567 | 2.3 | chain3 |
| 55 | 0 | 3 | 7688567 | 1.83 | chain3 |
| 60 | 1 | 4 | 4566 | 1.2 | chain3 |
| 60 | 0 | 5 | 34533 | 0.7 | chain3 |
| 75 | 0 | 8 | 987689 | 0.8 | chain3 |
| 90 | 1 | 9 | 3453454 | 1.18 | chain3 |
| 50 | 1 | 9 | 567867 | 1.1 | chain3 |
| 65 | 1 | 9 | 565567 | 1.1 | chain3 |
| 82 | 0 | 2 | 34345 | 1.3 | chain3 |
| 82 | 1 | 2 | 754567 | 0.6 | chain3 |
| 50 | 0 | 3 | 745676 | 2.9 | chain3 |
| 70 | 1 | 4 | 3453454 | 1.3 | chain3 |
| 50 | 0 | 4 | 567867 | 1.18 | chain3 |
| 65 | 0 | 8 | 5678 | 1 | chain3 |
| ... | ... | .. | ... | ... | ... |

The pivot table for orders and transactions to track the chains being used for transactions is given in Table 8.3.

**TABLE 8.3**
**Pivot Table for Comparison**

| Pivot Table for Orders | Pivot Table for Transactions |
|---|---|
| **Row Labels** | **Row Labels** |
| **23** | **23** |
|   **0.8** |   **0.8** |
|     chain8 |     chain8 |
| **47** | **45** |
|   **1** |   **0.7** |
|     chain6 |     chain1 |
|   **1.5** |     chain2 |
|     chain4 |     chain6 |
|   **6.1** |     chain8 |
|     chain7 |   **1** |
| **52** |     chain1 |
|   **2.9** |     chain5 |
|     chain5 | **47** |
| **53** | **1** |
|   **1** |     chain6 |
|     chain1 | **52** |
| **55** |   **2.9** |
|   **0.7** |     chain5 |
|     chain4 | **60** |
| **59** |   **1.18** |
|   **2.9** |     chain5 |
|     chain2 |   **1.3** |
| **60** |     chain5 |
|   **0.9** | **63** |
|     chain8 |   **1.2** |
|   **1.18** |     chain8 |
|     chain5 | **70** |
|   **1.3** |   **1.2** |
|     chain5 |     chain6 |
| **61** | **80** |
|   **1** |   **1.3** |
|     chain4 |     chain7 |
| **63** | **81** |
|   **1.2** | **2** |
|     chain8 |     chain3 |
| **66** | **91** |
|   **1.1** |   **1.3** |
|     chain3 |     chain4 |
| **68** | **94** |
|   **1.1** |   **0.9** |
|     chain2 |     chain3 |
|   **1.2** | **111** |
|     chain2 |   **1.1** |

(*continued*)

**TABLE 8.3  (Continued)**
**Pivot Table for Comparison**

| Pivot Table for Orders | Pivot Table for Transactions |
|---|---|
| 2 | chain4 |
| chain8 | 112 |
| 69 | 0.9 |
| 1.2 | chain3 |
| chain6 | 122 |
| 1.3 | 1 |
| chain1 | chain4 |
| 70 | 123 |
| 1.2 | 1.18 |
| chain6 | chain2 |
| 75 | 124 |
| 1.18 | 1.1 |
| chain1 | chain2 |
| 76 | 125 |
| 0.8 | 1 |
| chain3 | chain3 |
| 78 | 127 |
| 0.7 | 0.7 |
| chain6 | chain4 |
| 80 | 128 |
| 1.3 | 1.3 |
| chain7 | chain6 |
| 81 | 129 |
| 2 | 0.7 |
| chain3 | chain2 |
| 84 | 146 |
| 2.7 | 0.6 |
| chain1 | chain3 |
| 91 | 148 |
| 1.3 | 0.8 |
| chain4 | chain2 |
| 92 | 149 |
| 1 | 2.3 |
| chain5 | chain1 |
| 94 | 157 |
| 0.9 | 1 |
| chain3 | chain1 |
| 96 | 159 |
| 0.8 | 0.6 |
| chain3 | chain2 |
| 1 | 160 |
| chain5 | 1.3 |
| 102 | chain5 |
| 0.7 | 168 |
| chain6 | 1.1 |

## TABLE 8.3 (Continued)
## Pivot Table for Comparison

| Pivot Table for Orders | Pivot Table for Transactions |
|---|---|
| **0.75** | chain6 |
| chain6 | **220** |
| **109** | **1** |
| **0.9** | chain7 |
| chain7 | **231** |
| **110** | **2.7** |
| **1.6** | chain4 |
| chain4 | **235** |
| **111** | **1** |
| **1.1** | chain8 |

Further, the matching of pivot tables based on the chains being used for transactions is given in Table 8.4.

## TABLE 8.4
## Matching of Chains in Orders and Transactions Tables

| Pivot Table for Order | Pivot Table for Transactions |
|---|---|
| **Row Labels** | **Row Labels** |
| **23** | **23** |
| **0.8** | **0.8** |
| chain8 | chain8 |
| **47** | **47** |
| **1** | **1** |
| chain6 | chain6 |
| **1.5** | **52** |
| chain4 | **2.9** |
| **6.1** | chain5 |
| chain7 | **60** |
| **52** | **1.18** |
| **2.9** | chain5 |
| chain5 | **1.3** |
| **60** | chain5 |
| **0.9** | **149** |
| chain8 | **2.3** |
| **1.18** | chain1 |
| chain5 | **582** |
| **1.3** | **0.7** |
| chain5 | chain1 |

(*continued*)

**TABLE 8.4 (Continued)**
**Matching of Chains in Orders and Transactions Tables**

| Pivot Table for Order | Pivot Table for Transactions |
|---|---|
| **149** | chain7 |
| **2.3** | **0.8** |
| chain1 | chain6 |
| **582** | **1.1** |
| **0.7** | chain2 |
| chain1 | **1.18** |
| chain7 | chain4 |
| chain8 | chain8 |
| **0.8** | **1.7** |
| chain6 | chain4 |
| chain8 | **1.83** |
| **1** | chain1 |
| chain4 | **2** |
| chain5 | chain1 |
| **1.1** | **Grand Total** |
| chain2 | |
| chain8 | |
| **1.18** | |
| chain4 | |
| chain5 | |
| chain8 | |
| **1.3** | |
| chain1 | |
| **1.7** | |
| chain4 | |
| **1.83** | |
| chain1 | |
| chain3 | |
| chain6 | |
| **2** | |
| chain1 | |
| **2.1** | |
| chain2 | |

The success rate is calculated based on the coloring of matching between the same chains being used for the same account id. It is given in Table 8.5.

**TABLE 8.5**
**Secured Transfer of Data with or without Any Data Leakage**

| Account Id in Order | Account Id in Transactions | Blockchain # in Order | Blockchain # in Transaction | Success/ Failure | % of Success Rate |
|---|---|---|---|---|---|
| 23 | 23 | Chain 8 | Chain 8 | Success | 100 |
| 47 | 47 | Chains 6, 4, 7 | Chain 6 | Partial | 35 |
| 52 | 52 | Chain 5 | Chain 5 | Success | 100 |
| 60 | 60 | Chain 8, 5, 5 | Chain 5, 5 | Partial | 70 |
| 149 | 149 | Chain 1 | Chain 1 | Success | 100 |
| 582 | 582 | Chain 1, 7, 8, 6,8,4,5,2,8,4,5 …. (19 Chains) | Chain 1,7,6,2,4,8…. (9 Chains) | Partial | 50 |

## 8.9   RESULTS ANALYSIS AND DISCUSSION

The analysis performed over the actual transactions and orders made gives insight into the secured data transfer-based information. Table 8.5 describes the success rate of the transactions being made.

From the above table, it is understood that some transactions do not use the same Blockchain and thus lead to data leakage. This 100% safe transfer with the same chain id with the same evidence can be achieved if AI/ML algorithms are employed to track and monitor the chain usage during the transactions. The data is projected in Figure 8.1. From the figure, it is inferred that the transactions that used Blockchain with AI can provide a success rate of 1, whereas the others provide less rate.

## 8.10   CONCLUSION

Blockchain is an important step forward in digital security that will assist to ensure that the CIA triads of cybersecurity are implemented. However, there may be some challenges in actually using it due to the complexity of its implementation. In this section, we apply Hyperledger fabric to a dataset in order to analyze it. Included in this framework is a container technology for hosting the chain code, or smart contracts, that specifies the operational parameters of the network. The complexity of the global economy is taken into account, and the system is built to support a wide variety of pluggable components. This is helpful for specialized transactions, such as a transfer of ownership of an asset. Hyperledger fabric outperforms public Blockchain because it is a private Blockchain network; hence, transactions on this network do not need validation. However, dataset-based research reveals that those changes are occurring

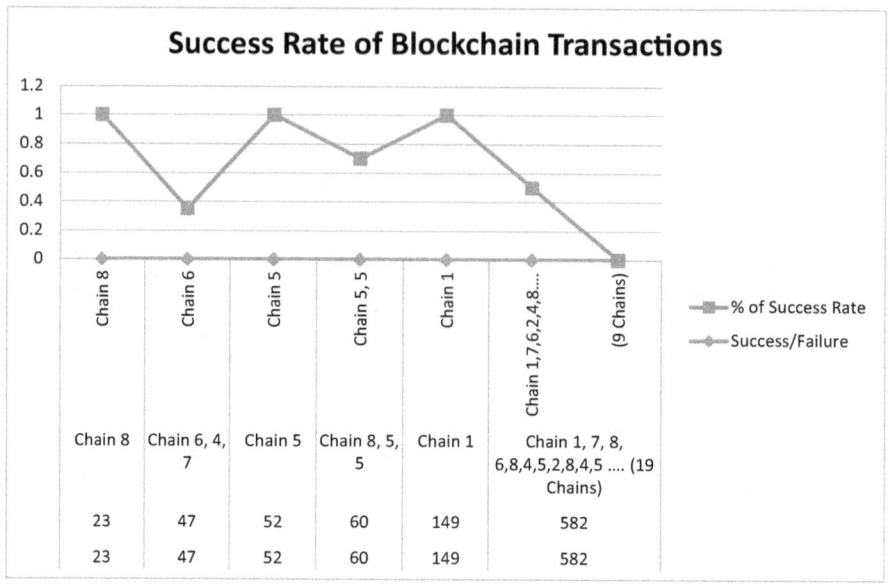

**FIGURE 8.1**    Success rate of Blockchain transactions.

in some chain ids and that some transactions were not fully completed when they depended on one another. To ensure that transactions are completed without data leakage or security breaches, it is clear that some intelligence-based monitoring and tracking mechanism is required, even though Blockchain provides a safe means of data transmission.

AI gives immediate insights to guide us through thousands of daily warnings and reduces response times by aggregating threat intelligence from millions of academic papers, blogs, and news stories. Digital technology use is harmful. In digital terms, the complexity of existing IT systems, lack of expertise, or unclear danger still impact us directly. Whatever the cause, cybersecurity is as important as real-world safety precautions.

All vital data, such as identification, money, and private information, can be hacked, yet Internet data security is inadequate. Blockchain and AI are born from predictive, reactive, and validating human activities, making them viable security safeguards.

## REFERENCES

Abu-elezz, Israa, Asma Hassan, Anjanarani Nazeemudeen, Mowafa Househ, and Alaa Abd-alrazaq, The benefits and threats of blockchain technology in healthcare: A scoping review, *International Journal of Medical Informatics*, Volume 142, 2020, 104246.

Clim, Antonio, Răzvan Daniel Zota, and Radu Constantinescu, Data exchanges based on blockchain in m-health applications, *Procedia Computer Science*, Volume 160, 2019, 281–288.

Dinh, Thang N. and My T. Thai, AI and blockchain: A disruptive integration, *Computer*, Volume 51(9), 2018, 48–53.

Haleem, Abid, Mohd Javaid, Ravi Pratap Singh, Rajiv Suman, and Shanay Rab, Blockchain technology applications in healthcare: An overview, *International Journal of Intelligent Networks*, Volume 2, 2021, 130–139.

Hylock, Ray Hales, and Zeng Xiaoming, A blockchain framework for patient-centered health records and exchange (HealthChain): Evaluation and proof-of-concept study. *Journal of Medical Internet Research*, Volume 21(8), 2019 Aug 31, e13592. doi: 10.2196/13592. PMID: 31471959; PMCID: PMC6743266

Landi, Heather. Healthcare Data Breaches Hit All-Time High in 2021, Impacting 45M People. *Fierce Healthcare*, February 1, 2022. www.fiercehealthcare.com/health-tech/healthc are-data-breaches-hit-all-time-high-2021-impacting-45m-people

Mathew A.R., Cyber Security through Blockchain Technology, *International Journal of Engineering and Advanced Technology (IJEAT)*, Volume 9(1), 2019, 382–3824.

Sharma, Yogesh, and B. Balamurugan, Preserving the privacy of electronic health records using blockchain, *Procedia Computer Science*, Volume 173, 2020, 171–180.

Shi, Shuyun, Debiao He, Li Li, Neeraj Kumar, Muhammad Khurram Khan, and Kim-Kwang Raymond Choo, Applications of blockchain in ensuring the security and privacy of electronic health record systems: A survey, *Computers & Security*, Volume 97, 2020, 101966.

Tariq, Noshina, Ayesha Qamar, Muhammad Asim, and Farrukh Aslam Khan, Blockchain and smart healthcare security: Asurvey, *Procedia Computer Science*, Volume 175, 2020, 615–620.

Xiaohua, Feng, Conrad Marc, Eze Elias, and Hussein Khalid, Artificial intelligence and blockchain for future cyber security application, *2021 IEEE Intl Conf on Dependable, Autonomous and Secure Computing, Intl Conf on Pervasive Intelligence and Computing, Intl Conf on Cloud and Big Data Computing, Intl Conf on Cyber Science and Technology Congress (DASC/PiCom/CBDCom/CyberSciTech)*, 2021,802–805.doi: 10.1109/ DASC-PICom-CBDCom-CyberSciTech52372.2021.00133

# 9 Implementation of a Blockchain-Based Secure Cloud Computing Mechanism for Transactions

*Muhammad Zunnurain Hussain,[1]*
*Muhammad Zulkifl Hasan,[2] Adnan Nabeel*
*Qureshi,[3] and Ghulam Mustafa[4]*

[1] Department of Computer Science, Bahria University Lahore Campus, Lahore, Pakistan

[2] Faculty of Computer Science and Information Technology, University of Central Punjab, Lahore, Pakistan

[3] Department of Computer Science, Birmingham City University, United Kingdom

[4] Department of Computer Science, University of Central Punjab, Lahore, Pakistan

## 9.1 INTRODUCTION

In an era of data privacy and security, cloud computing has transformed data storage, access, and analysis. Centralized cloud computing raises data security, integrity, and trustworthiness risks. Blockchain may address these difficulties. Blockchain is now a secure, decentralized platform for transactions. Blockchain's immutability, transparency, and distributed consensus make cloud computing secure. This chapter examines transaction-safe blockchain cloud computing. Blockchain technology's security and transparency are combined with cloud computing's dynamic resource allocation and sharing. Blockchain-enabled cloud computing is beneficial. First, a tamper-proof, auditable ledger protects data and transactions. This safeguards data. Second, blockchain's decentralization eliminates single points of failure and makes the system more robust to attacks. Blockchain smart contracts automate transactional rules, avoiding human error. This blockchain-based secure cloud computing technology needs scalability, performance, and interoperability for implementation.

DOI: 10.1201/9781003407096-9

Design of a high-throughput consensus algorithm. Compatible, integrated cloud computing infrastructures ease adoption. Blockchain technology in cloud computing makes secure transactions. Integrating both technologies might change business, data storage, and privacy. This chapter will explore the technical details of the suggested implementation and its pros and cons.

### 9.1.1  RESEARCH AIMS

This project aims to demonstrate how blockchain technology may be employed in a dependable cloud computing infrastructure for digital currency transactions. The project uses blockchain technology to enhance cloud-based digital transactions' security, transparency, and efficiency. This chapter covers the deployment process to promote safe cloud computing in transactional situations.

The key goals of the research are:

- Cloud computing and distributed ledgers: overviews and sound bites.
- The blockchain and its potential cloud applications should be investigated.
- Understand the importance of blockchain protection for your cloud-based apps' data.
- Analyze contemporary strategies for guarding the confidentiality, veracity, and integrity of the public's data.

## 9.2  RESEARCH OBJECTIVES

This research examines blockchain technology's potential for cloud-based financial transactions. This research will help firms and regions embrace safe and transparent transactional practices by showing how to construct a secure cloud computing approach for transactions utilizing blockchain technology.

### 9.2.1  DESIGN A ROBUST ARCHITECTURE

Our study has two goals. A secure cloud computing architecture for transactional contexts must be designed and tested. This process involves identifying the relevant components, specifying how they interact, and creating a framework to assure the integrity and secrecy of transactions.

### 9.2.2  INTEGRATE BLOCKCHAIN TECHNOLOGY

In the second objective, blockchain technology will be integrated into a secure cloud computing mechanism to achieve the first objective. There is a need to examine various blockchain platforms to select the most appropriate for transactional purposes. Several steps are involved in the integration process, including designing smart contracts, implementing consensus mechanisms, ensuring data immutability, and ensuring decentralization of data.

### 9.2.3  Secure Transactions

A third objective of the project is to establish secure transactional mechanisms due to utilizing blockchain technology. As part of this process, methods will be developed to verify identity, secured communication channels will be established, and cryptographic techniques will be implemented to safeguard transactional data. By facilitating secure and trustworthy transactions within cloud computing environments, the aim is to facilitate the use of cloud computing.

### 9.2.4  Secure and Protect

Increased security should make cloud-based transactions safer and more private. Encryption, access control, and data protection may secure sensitive data. Secure cloud computing tackles privacy and security problems with standard cloud computing.

### 9.2.5  Performance and Scalability

This study concludes by assessing the secured cloud computing mechanism's performance and scalability. The system's performance, transaction processing speed, and capacity to handle more transactions may need to be tested. As organizations develop, the mechanism must be efficient and scalable to satisfy transaction demands.

## 9.3  BACKGROUND STUDY

### 9.3.1  Cloud Storage

1. Cloud storage stores data on internet-accessible servers. It simplifies and scales file storage, organization, and access from any internet-connected device. Large data center businesses provide cloud storage. These firms provide subscription-based storage for corporations and free or limited-capacity solutions for consumers.
2. Cloud storage is simple. PC, smartphone, and tablet users may access their data anytime, anyplace. Flexibility allows remote file exchange and collaboration.
3. Cloud storage is scalable. Users may increase data storage without hardware or infrastructure.
4. Cloud storage businesses secure data using data replication and backup. If a server or storage device fails, users' data is safe and accessible from other servers. Cloud storage recovers data after accidental loss, device failure, or natural disasters.
5. Cloud storage eliminates the cost of physical storage infrastructure. Consumers rent storage space instead. Pay-as-you-go reduces upfront costs and improves budgeting, especially for individuals and small businesses.
6. Cloud storage allows users to collaborate on shared data. Real-time communication, version control, and frictionless file sharing increase productivity and teamwork.

7. Cloud storage firms protect user data. Data is encrypted during transport and storage. Advanced authentication, access controls, and permissions management protect data.
8. Data Privacy: Remote data storage raises privacy concerns. Cloud storage customers should review privacy and security policies to protect their data.
9. Cloud storage requires a reliable internet connection. Unreliable connections impede data availability and productivity.
10. Vendor Lock-In: Switching cloud storage providers may be complicated and costly. Check data transfer and compatibility before picking a service.
11. Organisations must comply with industry-specific regulations while keeping sensitive data. Check data privacy laws before utilising cloud storage.
12. Cloud storage is simple, scalable, and economical. Remote servers and internet connections promote accessibility, collaboration, and data redundancy. Cloud storage providers must prioritise data security, privacy, and compliance. (Sengupta, Ruj, and Bit, n.d.)

### 9.3.2 Blockchain Overview

Blockchain is hot. New and extensively utilized, its significance has grown. Blockchain, founded in 2008, is changing communication, fees, and transaction tracking. Blockchain was designed as a distributed database. Blockchain technology may thrive because no central authority monitors and records transactions and member interactions. Mining machines with blockchain transaction copies cryptographically tag and validate each transacstion.

This technique offers safe, synchronized, and permanent timeshare registration. Everyone understands blockchain has software and business applications. Figure 9.1 depicts blockchain design (Shrimali and Patel, 2022).

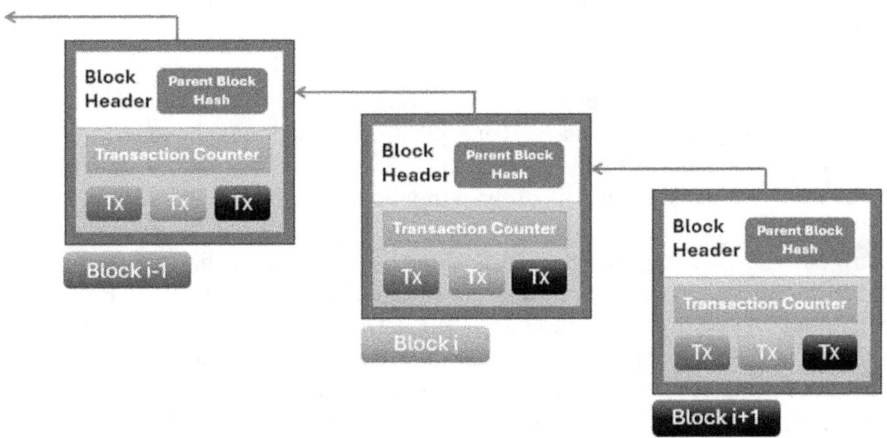

**FIGURE 9.1** Architecture of blockchain.

### 9.3.3 Blockchain Traits

After processing, transactions are broadcast to all nodes and organized into blocks. Blockchain's main goal was to simplify data exchange across networks. The blockchain's biggest strength is its decentralization. Blockchain networks use a consensus approach to verify data without a central authority or service fee (Wikipedia Contributors, 2022). Since every node stores transaction data, no central database or clearinghouse is required. This lowers the danger of a single catastrophic failure.

A consensus method is needed to ensure financial transactions are honest. The consensus algorithm avoids errors and maintains the blockchain's data. A transaction must also be confirmed by all parties. "Consent" characterizes this stage. Any blockchain member may send and receive transactions at that address, which is anonymous. Members will observe the encrypted dialogue, but the system will not reveal user information. Each blockchain transaction is time-stamped and digitally signed, so the organization may use that information to identify between blockchain participants. Thus, each block is inextricably linked.

#### 9.3.3.1 Decentralization

Blockchain operates on a decentralized network of computers, known as nodes, where no single central authority controls. This distributed nature eliminates the need for intermediaries and enables direct peer-to-peer transactions, enhancing transparency and reducing reliance on a single point of failure.

The characteristics of immutability and transparency are often cited in academic literature as key features of certain systems or technologies.

Upon the inclusion of data in a blockchain, the possibility of modifying or erasing the said data is rendered highly improbable. The property of immutability inherent in blockchain technology guarantees the preservation of data integrity, rendering it remarkably impervious to any attempts at unauthorized alteration. Furthermore, the inherent transparency of blockchain technology enables all involved parties to access and authenticate the transactions and information recorded on the distributed ledger.

##### 9.3.3.1.1 Security

Blockchain technology secures transactions and data using advanced encryption technologies. Cryptographic hashes connect each transaction to the preceding one to establish a safe chain of blocks. Decentralized consensus techniques like proof-of-work or proof-of-stake ensure blockchain transaction validation and addition.

Blockchain technology builds confidence by verifying transactions with many parties. A decentralized consensus process ensures that all nodes agree on transaction authenticity, eliminating the need for centralization. Blockchain technology's transparency and auditability let participants verify data integrity, building confidence. Blockchain technology removes middlemen and automates trust, improving efficiency and cost. Automated reconciliation eliminates manual reconciliation activities, decreasing transactional system costs. Blockchain-coded smart contracts may automate agreement enforcement. This may boost efficiency and save expenses.

### 9.3.3.2 Privacy and Control

While blockchain promotes transparency, it also provides privacy and control over data through cryptographic techniques. Participants can control their identities and choose to disclose specific information selectively. Private or permissioned blockchains can restrict access to authorized participants, ensuring the confidentiality and privacy of sensitive data.

### 9.3.3.3 Scalability

Scalability has challenged blockchain networks due to resource-intensive consensus mechanisms. However, advancements such as layer 2 solutions (e.g., Lightning Network) and sharding techniques are being developed to improve the scalability of blockchain networks, allowing them to handle larger transaction volumes and accommodate more users.

## 9.4 CLOUD COMPUTING TECHNIQUES

Cloud computing provides on-demand computer services. Without infrastructure, many software, storage, and processing power are available. Cloud computing manages resources. Three main cloud computing methods are the following:

IaaS: It is Cloud computing's base. Computing, networking, and storage resources online. User-managed software stack, OS, apps, and deployment. IaaS helps customers scale resources.

PaaS is vaguer. OSes, developer frameworks, and runtime environments are preconfigured. PaaS frees developers from infrastructure maintenance. Databases, communications queues, and authentication simplify development.

SaaS is user-centric cloud computing. Web browsers and client apps access full-featured applications online. SaaS users don't install or manage hardware. Email, CRM, collaboration, and document management are SaaS.

Cloud computing scales resources. Infrastructure is avoided.

Pay-as-you-go savings. Eliminating hardware upfront reduces cloud computing operating costs.

Any Internet-connected device may access cloud computing apps and data like remote work, communication, and resources.

Cloud providers back up data. Servers and data centers store data to prevent loss.

Cloud computing speeds software upgrades. Preconfigured environments and services expedite development.

Cloud computing benefits, but data security, vendor lock-in, and industry constraints must be addressed. Companies and individuals should assess their demands and choose a cloud computing method.

## 9.5 BLOCKCHAIN–CLOUD COMPUTING RESEARCH

This section shows how traditional cloud computing approaches have improved blockchain, cloud storage, and academic research safety.

### 9.5.1 Cloud Computing Methodology

Cloud storage, which distributes server space and IT assets, may benefit enterprises and individuals. Most scientific study improves Information and Communication Technology (ICT) and cloud service security. Security procedures include encryption, integrity checks, data reduplication, user lockout, auditing, and archiving. Cloud clients can detect data tampering via symmetric encryption and an encrypted bloom filter. Index-pointer-based tagging checks and creates processes. It monitors and gently removes troublemakers. It prioritizes client data and cloud provider exploitation. Random Response Method (RARE)-duplicating cloud data avoids response channels. Zero-knowledge users' data redundancy response structure shields side channels. Despite secure storage, integrity verification, effective user lockout, and data reduplication, data privacy is a nightmare. Most solutions are for small, unchanging data sets and not for cloud data security. Thus, studying blockchain-based cloud storage solutions benefits.

### 9.5.2 Cloud–Blockchain Research

The emergence of blockchain technology has sparked innovative frameworks and solutions across various sectors. Extensive research is underway to evaluate its applications in business operations, e-government systems, healthcare services, security protocols, resource sharing platforms, cloud computing, edge computing, and beyond. These studies aim to explore not only the potential benefits but also the challenges and future prospects associated with integrating blockchain into these diverse fields. All early investigations included blockchain technology and cloud storage security. Blockchain may help cloud computing and storage. We'll study blockchain's cloud applications. In our research our focus mainly lies around blockchain's cloud storage.

### 9.5.3 Blockchain Structure

Blockchain operates through a decentralized network where transactions are securely recorded and linked in a sequential chain of blocks. The architecture of blockchain is utilized in cloud, edge, and fog computing, merging with other extensive distributed models. It reinforces cloud security measures and traces its origins to the need for secure and transparent transactions. Cryptographic signatures play a pivotal role in authenticating and validating financial transactions within the blockchain network. Blockchain protects terminals, Trading eluded hackers. This simplifies finances (Coelho et al., 2022).

Zero-knowledge proof verifies blocks rapidly and confidentially. Zero-knowledge proof protects and speeds block data verification. Block data may be validated without chaining the transaction. The smart contract validates the zero-knowledge proof and root hash from transaction data. Zero-knowledge proof verification secures smart contract transactions and validates blocks quickly. The buffer compresses, modifies, and authorizes the transaction (Wang et al., 2023).

**Hash:** It's essential to blockchain's versatility. It cyphers block data.

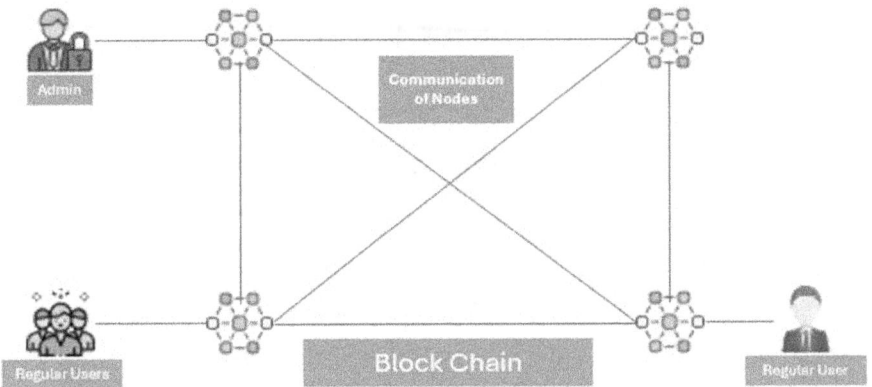

**FIGURE 9.2** Blockchain technology in P2P architecture.

Input may impact output. In Figure 9.2, many real-time applications employ Secure Hash Algorithm 256-bit (SHA-256). It calculates any number.

**Ledgers record business transactions.** Every node keeps a ledger of business transactions. Paper-and-pencil ledgers are utilized. New infotech uses this idea. Thus, centralized ledgers lose data and cannot independently verify transactions.

**Processed transactions.** This transaction index allows more process steps. Transaction intermediates are lost. Mining nodes update every transaction. Each block contains all transactions. Blockchain rejects invalid transactions automatically. This approach validates data rigidity since altering a block bit significantly alters the computed hash. For further security, each node retains the block hash. Since every node verifies the hash, this method is error-free and corruption-free. Merkle tree-compatible blockchains are listed below. Blockchain blocks include:

- The quantity of blocks or the vertical dimension of a block.
- The present hash value of the block.
- The hash value of the block that precedes the current one.
- The production of marijuana derived from Merkle tree roots is being discussed.
- Timestamp
- The block's measurements.
- This section comprises a roster of transactions. The following describes the operation of the blockchain: The vast majority of blockchain nodes are run autonomously.

Node communication is based on ledger data. Node registry problems degrade system performance. The blockchain executes user requests. One or more logs record all such transactions. As shown in Figure 9.3, this technique ensures blockchain immutability.

**Security of blockchain technology:** A business logic-related network. Blockchains use asymmetric key cryptography. Keys secure the transaction. Private keys validate transaction signatures. Public key controls private key signature.

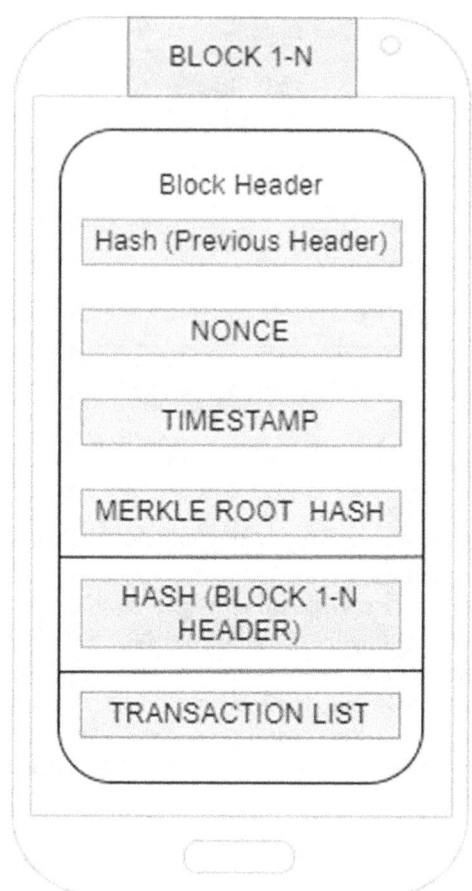

**FIGURE 9.3**  Generic blockchain model.

## 9.5.4 THE BITCOIN IDEA

Each bitcoin is a digital signature in a chain. Digital signatures transmit coins based on current and past hashed transactions. Depending on the property, the receiver may authenticate the signature. Figure 9.4 depicts the concept of Bitcoin. We explain Bitcoin here:

- An initial notice of any newly created transactions is sent to all nodes.
- New transactions are gathered by each node in the block individually.
- The weight must be distributed evenly over each block.
- When proof-of-work is discovered, a broadcast is sent to all of the other nodes that are part of the block.
- You must only engage in lawful transactions and avoid incurring any unnecessary costs.

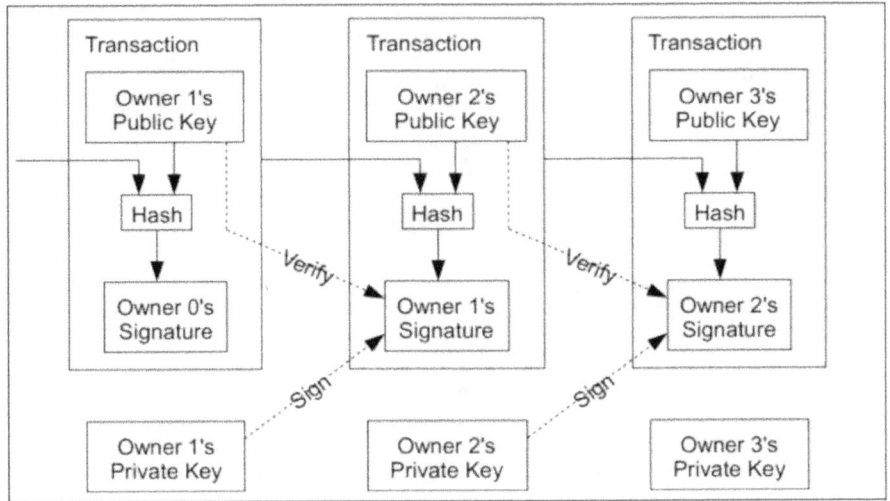

**FIGURE 9.4** Working process of Bitcoin.

If the hash function of the block is identical to that of the node, then the node will accept the block.

## 9.6 CLOUD AUTHENTICATION

Modern technology relies on application service providers like web and email servers. A corporation may provide several application services. The corporation may restrict service usage to qualified customers. Multiple authentication credentials are insecure, systemic, and administrative. Companies have encountered comparable cloud transition challenges. Numerous entities get these services, hence numerous protections are needed. Access permissions become harder and more expensive as services rise. Most programs improve system performance and corporate value. Thus, centralizing policy administration may improve authorization flexibility and scalability. Policy management simplifies policy modifications. Separating security and auditing duties makes the authorization system safer.

## 9.7 VARIABLES ASSOCIATED WITH BLOCKCHAIN SYSTEMS

The success of blockchain technology depends on many factors.

Cryptocurrency security: Digital money changed Internet trading. To restore the old system, just produce fresh paper and money. Electronic cash is most easy to use with credit cards. Distributors and salespeople must trust the system. Customers must request electronic money from bank or issuer before payment. Consumers save electronic cash on smart cards or other physical devices instead of online wallets.

These implementations might be traceable or anonymous. Before processing, a bank must review and double-check every transaction. Encryption and digital signatures

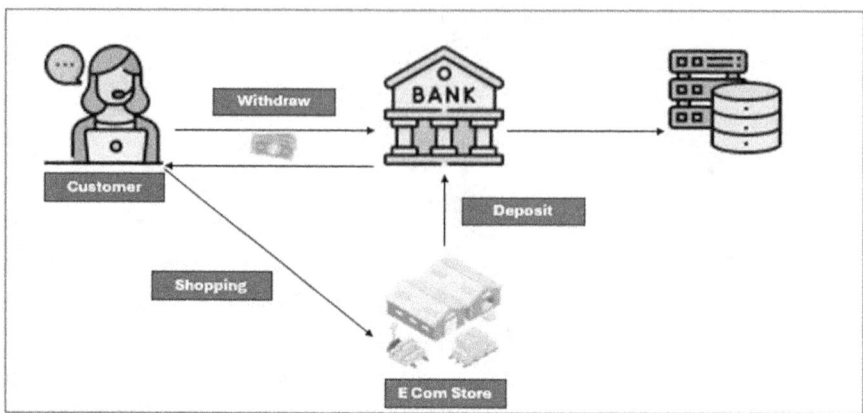

**FIGURE 9.5**   Workflow of e-cash.

secure electronic money communications in this application. The flow of e-cash in Figure 9.5 requires confidence between banks, consumers, and shops. Data, services, and information are significant resources. Access control protects stored data. For security, organizations require distinct access levels. Transferring access rights is important in various situations. A user's access credentials might be auctioned. In enterprises, one worker assigns another to utilize the same virtual computer to calculate.

## 9.8   SECURE CLOUD COMPUTING WITH BLOCKCHAIN

Cloud-based blockchains provide data preservation, replication, and transactional database access. Cloud blockchain transactions address these demands. Distributed ledger transactions increase blockchain transactions. Scalability and adaptability make cloud systems ideal for unexpected situations (Zheng et al., 2018). Remote datacenters encrypt. Correcting business transactions. Cloud customers may pick where to store and process data. System stability network node failure requires backup. Software and datacenter nodes replicate. For security, a blockchain-based distributed cloud must centrally map all software applications. Figure 9.6 illustrates a peer-to-peer based cloud architecture, where cloud clusters are connected via a peer-to-peer architectural model.

## 9.9   IMPLEMENTATIONS

### 9.9.1   IMMUTABILITY

SHA-256 cryptographic hashing records all blockchain transactions. The block header also hashes the previous block referenced by the linked block. Block data is hashed to create a fingerprint. Thus, blockchain blocks cannot be altered with current technology. Injections were Tx generic. Financial numbers may vary. The last Hash attribute stores the preceding block's hash. Merkle root and nonce are explained in previous chapters of the book.

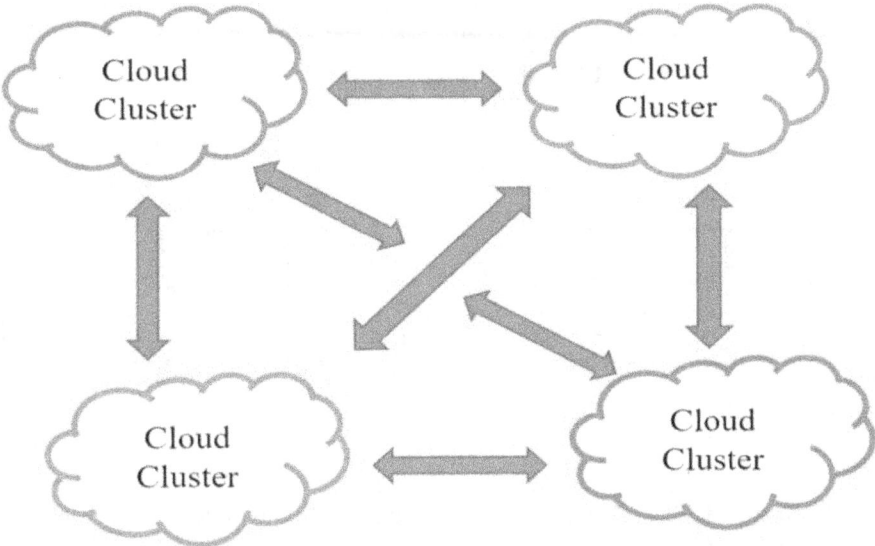

**FIGURE 9.6** P2P-based cloud architecture.

### 9.9.2 BLOCK HASH AND CHAIN

Each block has a hash. This hash concatenates the preceding block's hash, a SHA-256 hash obtained from it, and any additional block properties. Figure 9.7 displays the method from the Java class hash computation block. Transactions are serialized into JSON and appended to a block's attributes before hashing.

Chains are sequences of transactions or data records. The blockchain, or chain, secures and tamper-proofs all network transactions and data. Blockchain technology relies on the chain in numerous ways:

**Data Integrity:** The chain protects blockchain data. A cryptographic hash identifies and validates each chain block. The cryptographic hash of each block is derived using its contents and the hash of the preceding block in the chain.

**Consensus and Trust:** Decentralized network consensus relies on the chain. Proof-of-work or proof-of-stake consensus procedures guarantee that all network nodes agree on transaction legitimacy and chain order.

**Transaction History and Transparency:** Blockchain transactions are transparent and auditable. Each block has a timestamp and a reference to the preceding block, producing a continuous series of occurrences.

**Security and Tamper Resistance:** The chain's architecture and cryptography make it safe and tamper-resistant. The network's consensus procedures verify and add blocks securely.

### 9.9.3 ADDING TRANSACTIONS

Blockchain's decentralization is key. Append-only, they replicate data across the blockchain's distributed ledger. Like Bitcoin, nodes typically communicate

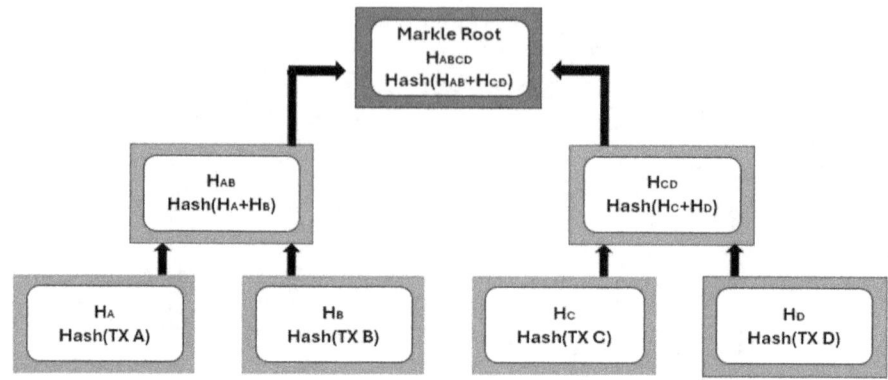

**FIGURE 9.7**   Merkle trees.

decentralizedly. HTTP APIs are decentralized blockchain implementations. That's another story. Transactions cover most everything. A transaction may comprise data or a smart contract to execute. A "smart contract" is a computer protocol that facilitates, verifies, or enforces a legally enforceable agreement (Alharby and van Moorsel, 2017; Atzei, Bartoletti, and Cimoli, 2017). Bitcoin transactions move Bitcoin amounts across accounts. Public keys and account IDs provide data confidentiality. Bitcoin only. A network stores transactions instead of a blockchain.

Blockchain consensus solves this. Verified consensus techniques and patterns are outside the scope of this chapter. There is mining consensus on Bitcoin blockchains. We'll examine this consensus later. The consensus method divides transactions into blocks and adds them to the chain. The chain verifies fresh transaction blocks before adding them. Merkle tree hashes are added to the block.

Merkle root hashes need a Merkle tree data structure. As shown in Figure 9.7 each block stores the Merkle toot, the hash of the root of the Merkle tree, a balanced binary tree of hashes where the inner nodes are hashes of the two-child hashes. This tree verifies block transactions. If any exchange character is changed, the Merkle root will be meaningless. A single fork of transaction hashes accumulates and verifies an entire block, enabling distributed block transmission. This calculates a block Merkle tree root. The Merkle tree unit test in the companion project appends a transaction to a block and checks that the Merkle roots have developed. The Merkle tree Java code is presented in Figure 9.8.

This unit test mimics checking and altering a block transaction outside the consensus process. This test simulates changing transaction data. Blockchain expansion is one-way. The blockchain's data structure is shared across nodes; therefore, blocks' data structures—including the Merkle toot—are hashed and connected. Every node may validate new blocks and display current block validity. Thus, a miner or node cannot construct a phoney block or change past transactions before the sun bursts into a supernova, giving everyone a great tan.

```java
import java.security.MessageDigest;
import java.security.NoSuchAlgorithmException;
import java.util.ArrayList;
import java.util.List;
public class MerkleTree {
    private List<String> transactions;
    private List<String> merkleTree;
    public MerkleTree(List<String> transactions) {
        this.transactions = transactions;
        this.merkleTree = buildMerkleTree();
    }
    private List<String> buildMerkleTree() {
        List<String> tree = new ArrayList<>(transactions);
        while (tree.size() > 1) {
            List<String> newTree = new ArrayList<>();
            for (int i = 0; i < tree.size(); i += 2) {
                String left = tree.get(i);
                String right = (i + 1 < tree.size()) ? tree.get(i + 1) : left;
                String parent = calculateHash(left + right);
                newTree.add(parent);
            }
            tree = newTree;
        }
        return tree;
    }
    private String calculate hash(String input) {
        try {
            MessageDigest digest = MessageDigest.getInstance("SHA-256");
            byte[] hash = digest.digest(input.getBytes());
            StringBuilder hex string = new StringBuilder();
            for (byte b: hash) {
                String hex = Integer.toHexString(0xff & b);
                if (hex.length() == 1)
                    hexString.append('0');
                hexString.append(hex);
            }
            Return hex string.toString();
        } catch (NoSuchAlgorithmException e) {
            e.printStackTrace();
        }
        return ";
    }
    public String getRoot() {
        return merkleTree.get(0);
    }

    public static void main(String[] args) {
        List<String> transactions = new ArrayList<>();
        transactions.add("Transaction1");
        transactions.add("Transaction2");
        transactions.add("Transaction3");
        transactions.add("Transaction4");
        MerkleTree merkleTree = new MerkleTree(transactions);
        System.out.println("Merkle Root: " + merkleTree.getRoot());
    }
}
```

**FIGURE 9.8**    Merkle tree implementation in JAVA.

### 9.9.4    DOCUMENTING WORK

"Mining" in Bitcoin involves combining transactions into a "block" and delivering it to the chain for validation. This is termed "consensus" in blockchain. Multiple consensus methods exist for dispersed networks. Public or permissioned blockchains govern how to use them. This essay covers blockchain mechanics, thus we'll utilize the proof-of-work consensus. Our white paper goes into greater detail. Mining nodes will solve a simple arithmetic problem while monitoring blockchain transactions. This puzzle creates a block hash with a set amount of leading zeros using a nonce value that changes with each iteration until the hash is found.

## 9.10   BLOCKCHAIN-BASED CRYPTOGRAPHIC WALLET DECRYPTION

The utilization of cryptographic wallets is imperative in ensuring the security of digital assets in blockchain systems. The security of sensitive data in these wallets is ensured through the utilization of private keys and digital signatures. The decryption of the contents of a cryptographic wallet is necessary. The utilization of blockchain technology has the potential to enhance the security of wallet decryption.

**The process of securing a wallet through the use of encryption techniques:** The mention of wallet encryption should precede that of decryption. The wallet software generates a pair of cryptographic keys, namely, a public key that is utilized for receiving funds and a private key that is employed for validating transactions. The encrypted storage of keys and wallet information is facilitated by the blockchain through the implementation of robust cryptographic methods.

**The process of decrypting data:** The decryption procedure entails the retrieval of encrypted wallet data from the blockchain, followed by the decryption of the said data to gain access to private keys and other confidential information.

**Identification of a wallet:** The user specifies the target wallet for decryption within the blockchain network. This task can be accomplished by furnishing the corresponding wallet address or any other distinct identifier.

**The process of gaining entry to encrypted wallet data:** Upon receipt of the user's decryption request, the blockchain system initiates a search for the encrypted wallet data linked to the designated wallet identifier. The aforementioned information is commonly stored in a predetermined storage site within the blockchain network, which may include a smart contract or a particular transaction.

**The process of verifying and authorizing:** In order to establish the user's ownership of the wallet, it may be necessary to implement supplementary verification measures during the decryption procedure. This may involve verifying ownership by means of digital signatures or multi-factor authentication mechanisms. The credentials of the user are validated by the blockchain network in order to establish their authorization.

**The process of obtaining an encryption key:** Upon successful confirmation of the user's authorization, the decryption procedure proceeds to obtain the encryption key(s) necessary for the decryption of the wallet. The storage of these keys could potentially be within the blockchain or linked to the user's identity or account.

**Decryption:** The wallet software decrypts encrypted wallet data by utilizing the designated encryption key(s), thereby granting the user access to confidential keys, digital signatures, and additional data pertaining to the blockchain network.

The utilization of blockchain technology's decentralized and secure nature in cryptographic wallet decryption has the potential to enhance wallet security. The wallet contents are safeguarded through the implementation of access controls, which

**TABLE 9.1**
**Brief Overview of Literature**

| No | Author | Methodology | Gaps | Results |
|---|---|---|---|---|
| 1 | Rahman et al. (2023) | This study proposes a methodical methodology for constructing a secure architecture based on blockchain-software-defined network (SDN) technology for cloud computing within the framework of intelligent industrial IoT. | This article aims to address a research gap pertaining to the absence of a comprehensive and secure architecture that is specifically tailored for cloud computing. | The present study proposes a secure architecture for cloud computing that is based on blockchain and software-defined networking (SDN). |
| 2 | Mahajan et al. (2023) | This study proposes a methodical strategy for the assimilation of Healthcare 4.0 and blockchain technologies into cloud-based electronic health record (EHR) systems with enhanced security measures. | There is a requirement for a unified and safeguarded resolution that amalgamates Healthcare 4.0 and blockchain technologies in cloud-based electronic health record (EHR) systems. | This study aims to showcase the advantages and efficacy of the amalgamation of Healthcare 4.0 and blockchain technologies into cloud-based electronic health record (EHR) systems, with a focus on enhancing security measures. |
| 3 | Yu et al. (2023) | This study focuses on the development and execution of a blockchain-powered authentication and authorization system for distributed mobile cloud computing services. | Many current authentication schemes are dependent on centralized servers for authentication purposes, which can potentially create security weaknesses and a single point of failure. | This study aims to showcase the efficacy and advantages of the proposed authentication and authorization mechanism based on blockchain technology in the context of distributed mobile cloud computing services. |

(continued)

**TABLE 9.1 (Continued)**
**Brief Overview of Literature**

| No | Author | Methodology | Gaps | Results |
|----|--------|-------------|------|---------|
| 4 | Murala et al. (2023) | The implementation of a blockchain-based technology for a cloud-based electronic health record (EHR) system is proposed to enhance its security. | Systems frequently encounter obstacles pertaining to data confidentiality, breaches of security, and absence of confidence in centralized systems. | The present study aims to investigate the feasibility of implementing a secure electronic health record (EHR) system within a cloud-based environment, utilizing blockchain technology. |
| 5 | Khezr et al. (2023) | This study aims to explore a framework that ensures the security and reliability of monetizing Internet of Things (IoT) data. The proposed framework leverages the use of blockchain and fog computing technologies. | There is a pressing need within the realm of Internet of Things (IoT) to establish a framework that is both secure and dependable for the purpose of monetizing data. This framework must also take into account the challenges that are associated with data security, trust, and privacy. | This study aims to evaluate the efficacy and advantages of a proposed framework designed for the secure and reliable monetization of Internet of Things (IoT) data. The framework leverages the combined capabilities of blockchain and fog computing technologies. |
| 6 | Gong and Navimipour (2022) | Undertaking a comprehensive and methodical examination of scholarly literature pertaining to blockchain-based methodologies utilized in the context of cloud computing. | The implications within the realm of cloud computing are currently undergoing a process of evolution. | Various implementation strategies for integrating blockchain technology with cloud computing have been proposed, including the utilization of private and permissioned blockchains, as well as hybrid cloud-blockchain architectures. |

| 7 | Habib et al. (2022) | A methodical approach was devised to systematically retrieve pertinent literature from diverse sources, such as scholarly databases, research papers, conference proceedings, and reputable periodicals. | The article presents a comprehensive analysis of blockchain technology, encompassing its advantages, obstacles, use cases, and incorporation with cloud computing. | Habib et al. (2022) provides a comprehensive analysis of blockchain technology, encompassing its advantages, obstacles, use cases, and incorporation with cloud computing. |
| 8 | Zhang et al. (2022) | This study aims to examine a knowledge-based system designed for a cognitive cloud computing model based on blockchain technology, with a specific focus on enhancing security measures. | There is a requirement for an enhanced security mechanism in cloud computing environments that utilizes the advantages of blockchain technology and cognitive computing. | The present study aims to showcase the efficacy and benefits of the suggested knowledge-driven system in the context of a cognitive cloud computing model based on blockchain technology, with a specific focus on enhancing security measures. |
| 9 | Prabakaran and Ramachandran (2022) | This study focuses on the development and deployment of a secure communication channel for financial transactions within a cloud-based environment. The proposed solution leverages blockchain technology to ensure the integrity and confidentiality of sensitive financial data. | The challenge of maintaining the security and reliability of said transactions persists. | This study aims to evaluate the efficacy of the proposed secure channel for financial transactions within a cloud-based setting utilizing blockchain technology. The system has the capability to manage a significant volume of simultaneous transactions. |
| 10 | Krishnaraj et al. (2022) | The present study undertook an analysis of the available literature and information to discern the primary obstacles encountered by conventional cloud computing models and the potential advantages offered by decentralized cloud/fog solutions based on blockchain technology. | The objective is to investigate the prospective advancements in cloud computing, with particular emphasis on decentralized cloud/fog solutions based on blockchain technology. As cloud computing has gained widespread adoption, there has been an increasing inclination toward utilizing blockchain technology. | The traditional cloud computing models are confronted with various challenges such as apprehensions regarding data security, privacy, scalability, vendor lock-in, centralized control, and single points of failure. |

*(continued)*

**TABLE 9.1 (Continued)**
**Brief Overview of Literature**

| No | Author | Methodology | Gaps | Results |
|----|--------|-------------|------|---------|
| 11 | Kumar et al. (2019) | The statistical approach known as the Power of the Test (POW). The consensus technique is employed in the domains of cloud and fog computing. Moreover, it employs the expectation-maximization algorithm and employing factorization techniques on the polynomial matrix. | The process of retrieving the data is time-consuming. The delay in obtaining improved consent can be attributed to the presence of misclassified data blocks. | The convergence time for the solutions is reduced and the mathematical models are configured more efficiently. |
| 12 | Banerjee et al. (2018) | This study examined IoT data sets and blockchain mechanism impacts from 2016 to the present. | They identify dangers to blockchain's future. Data intervention improves. Hardware and software are vulnerable. | They addressed public network data dependability and dissemination. |
| 13 | Jiao et al. (2019) | This study presented a resource-efficient auction-based market model. Management was decentralized. The continuous demand and multiple demand supply systems maximize well-being. | Blockchain developers limit marginal earnings as mining grows. Block streaming prevented double-spend attacks. | An optimum winner selection solution is found. The miner, rationality, and guaranteed lower limit determine it. |

restrict unauthorized users from accessing and deciphering private keys and other confidential information.

## 9.11  LITERATURE SURVEY

Understanding the extent of the investigation is made more accessible by identifying the knowledge gap. There is a tabular presentation of the complete survey in Table 9.1.

## 9.12  CONCLUSION

Blockchain, a popular financial technology, may enable digital monetary transactions across different nodes. Blockchain data is kept on patrons' Peer to Peer (P2P) networks, maximizing computing resources. "Proof of Work" and "Proof of Rights and Interests" are the most used blockchain transaction security methods. This chapter discusses blockchain technology and cloud computing. We'll discuss blockchain networks and cloud infrastructures. This unified system prioritizes data server–consumer–security trust. Blockchain history and merits and downsides were examined.

To identify the challenges of this integration, we studied the literature and data. The report shows that blockchain-based cloud solutions are still being studied. Researchers face several obstacles, including access restrictions. Sharing data for financial benefit disrupts the network and causes unexpected financial losses. Fake accounts reduce system scalability. Future models should solve the aforesaid difficulties.

## REFERENCES

Alharby, Maher, and Aad van Moorsel. 2017. "Blockchain-Based Smart Contracts: A Systematic Mapping Study." arXiv preprint arXiv:1710.06372

Atzei, Nicola, Massimo Bartoletti, and Tiziana Cimoli. 2017. "A Survey of Attacks on Ethereum Smart Contracts (SoK)." In *Lecture Notes in Computer Science*, 164–86. Berlin, Heidelberg: Springer Berlin Heidelberg.

Banerjee, Mandrita, Junghee Lee, and Kim-Kwang Raymond Choo. 2018. "A Blockchain Future for Internet of Things Security: A Position." *Digit. Commun. Netw.* 4(3): 149–60.

Coelho, Raiane, Regina Braga, José Maria N David, Victor Stroele, Fernanda Campos, and Mário Dantas. 2022. "A Blockchain-Based Architecture for Trust in Collaborative Experimentation." *J. Grid Comput.* 20(4): 35.

Gong, Jianhu, and Nima Jafari Navimipour. 2022. "An In-Depth and Systematic Literature Review on the Blockchain-Based Approaches for Cloud Computing." *Cluster Computing.* 25(1): 383–400.

Habib, Gousia, Spash Sharma, Sara Ibrahim, Imtiaz Ahmad, Shaima Qureshi, and Malik Ishfaq. 2022. "Blockchain Technology: Benefits, Challenges, Applications, and Integration of Blockchain Technology with Cloud Computing." *Future Internet.* 14(11): 341.

Jiao, Yutao, Ping Wang, Dusit Niyato, and Kongrath Suankaewmanee. 2019. "Auction Mechanisms in Cloud/Fog Computing Resource Allocation for Public Blockchain Networks." *IEEE Trans. Parallel Distrib. Syst.* 30(9): 1975–89.

Khezr, Seyednima, Abdulsalam Yassine, and Rachid Benlamri. 2023. "Towards a Secure and Dependable IoT Data Monetization Using Blockchain and Fog Computing." *Cluster Computing.* 26(2): 1551–64.

Krishnaraj, N., K. Bellam, B. Sivakumar, and A. Daniel, 2022. "The Future of Cloud Computing: Blockchain-Based Decentralized Cloud/Fog Solutions–Challenges, Opportunities, and Standards." *Blockchain Security in Cloud Computing*. 207–26.

Kumar, Gulshan, Rahul Saha, Mritunjay Kumar Rai, Reji Thomas, and Tai-Hoon Kim. 2019. "Proof-of-Work Consensus Approach in Blockchain Technology for Cloud and Fog Computing Using Maximization-Factorization." *IEEE Internet Things J.* 6(4): 6835–42.

Landi, Heather. "Healthcare Data Breaches Hit All-Time High in 2021, Impacting 45M People." *Fierce Healthcare*, February 1, 2022. www.fiercehealthcare.com/health-tech/healthcare-data-breaches-hit-all-time-high-2021-impacting-45m-people.

Mahajan, Hemant B., Ameer Sardar Rashid, Aparna A Junnarkar, Nilesh Uke, Sarita D Deshpande, Pravin R Futane, Ahmed Alkhayyat, and Bilal Alhayani. 2023. "Integration of Healthcare 4.0 and Blockchain into Secure Cloud-Based Electronic Health Records Systems." *Applied Nanoscience*. 13(3): 2329–42.

Panda, Sandeep Kumar and Mishra, Vaibhav and Dash, Sujata Priyambada and Pani, Ashis Kumar. 2023. "ecuring Electronic Health Record System in Cloud Environment Using Blockchain Technology." *Recent Advances in Blockchain Technology: Real-World Applications* (pp. 89–116). Cham: Springer International Publishing.

Prabakaran, D. and Shyamala Ramachandran, 2022. *Secure Channel for Financial Transactions in a Cloud Environment Using Blockchain Technology*.

Rahman, Anichur, Md Jaihul Islam, Shahab S. Band, Ghulam Muhammad, Kamrul Hasan, and Prayag Tiwari. 2023. "Towards a Blockchain-SDN-Based Secure Architecture for Cloud Computing in Smart Industrial IoT." *Digit. Commun. Netw.* 9 (2): 411–421.

Sengupta, Jayasree, Sushmita Ruj, and Sipra Das Bit. n.d. "Blockchain-Enabled Verifiable Collaborative Learning for Industrial IoT." *Ieee.Org*.

Shrimali, Bela, and Hiren B. Patel. 2022. "Blockchain State-of-the-Art: Architecture, Use Cases, Consensus, Challenges, and Opportunities." *J. King Saud Univ.—Comput. Inf. Sci.* 34(9): 6793–6807.

Wang, Jin, Wei Ou, Osama Alfarraj, Amr Tolba, Gwang-Jun Kim, and Yongjun Ren. 2023. "Block Verification Mechanism Based on Zero-Knowledge Proof In." *Comput. Syst. Sci. Eng.* 45(2): 1805–19.

Wikipedia Contributors. 2022. "Bandaranaike Airport Attack." *Wikipedia, The Free Encyclopedia*.

Yu, Linsheng, Mingxing He, Hongbin Liang, Ling Xiong, and Yang Liu. 2023. "A Blockchain-Based Authentication and Authorization Scheme for Distributed Mobile Cloud Computing Services." *Sensors*. 23(3): 1264.

Zhang, Honglei, Zhenbo Zang, and BalaAnand Muthu. 2022. "Knowledge-Based Systems for Blockchain-Based Cognitive Cloud Computing Model for Security Purposes." IJMSSC. 13(04): 2241002.

Zheng, Zibin, Shaoan Xie, Hong Ning Dai, Xiangping Chen, and Huaimin Wang. 2018. "Blockchain Challenges and Opportunities: A Survey." *Int. J. Web Grid Serv.* 14(4): 352.

# 10 Geolocation-Based Smart Land Registry Process with Privacy Preservation Using Blockchain Technology and IPFS

*Narendra Kumar Dewangan and
Preeti Chandrakar*

Department of Computer Science & Engineering, National
Institute of Technology Raipur, India

## 10.1 INTRODUCTION

In e-governance services, land registry is an integral part. The government takes some charges and provides this service through the local authority, which updates the registry records in the ledger. Traditionally, these records are maintained on centralized servers, which have many vulnerabilities and risks. In the online and application-based registry systems, the records are entered by individuals. The approval authority takes the data, verifies the data through the documents available in the hard copies and previous records, and approves the registration of the land. When a situation like coronavirus disease 2019 surfaces, the registration process for the land and properties is affected. The real estate business is one of the most important business models for any country, and when it is down, the economy of that country is also affected. The traditional system of land registration limits the globalization of trust between the parties. The deniable conditions of the parties in the land registry process are a common problem in the current land registry system in India.

Blockchain is a distributed, decentralized, immutable, transparent, secure, and fast system to store data. Devices like the Internet of Things (IoT) are used for the recording of real-time data. Blockchains can store real-time data in distributed storage. This storage is free from single points of failure because of the availability of data at any point. Blockchain data, once placed in a block, cannot be changed or deleted. Altering data is not possible on the blockchain. In blockchain, each block is connected to its previous block by the cryptographic hash pointer. This hash is known as the previous hash in the blockchain's structure. This part defines the structure and workings of the blockchain, cloud storage, and connectivity of IoT devices with the cloud. At the end of the introduction part, the goals of the chapter are defined, along with the organization of the chapter.

Blocks in the blockchain consist of the data structure known as the Markle tree. This tree holds the list of transactions in a binary tree format. The transaction is converted into cryptographic hash form before compiling on the blockchain. The root of the block, known as the Markle root, is in charge of this transaction. Other block structures included the timestamp for the recording of the block creation time and verification process in future transactions. Block headers consist of the previous hash (the block header of the previous block), nonce, and block index. Nonce is defined as the number only used once in any blockchain for identification of the block's uniqueness. All this data is compressed with one hash known as the block hash. These block hashes are used to point to the next and previous blocks. One block is connected to another, making a chain. This structure is known as the blockchain. The Interplanetary File System (IPFS) is used as the distributed storage for large files and can store a file with a different name but the same hash code. A file may supply a different name in the same system but return only one hash value. A tree structure of the file is maintained on all connected and updated nodes of the IPFS. The authors of [1] described the use of blockchain and IoT devices to implement the cold chain system. Temperature and location data are gathered by the gadget and sent to the crucial blockchain infrastructure, Hyperledger Sawtooth. The application's goal is to let customers follow products through a supply chain and alert the driver when a container's temperature is outside of the acceptable range. We interconnected the application, the devices, and the blockchain database in order to store the shipment log as a blockchain transaction. No one has the authority to alter, edit, or delete any information; therefore, this idea can guarantee data integrity, trustworthiness, and traceability. Vistro et al. [2] undertook a systematic literature analysis to analyze and synthesize all available information on the application of innovative blockchain technology to the agriculture and food supply chains. In this research, the real-time use of blockchain in agriculture supply chain management was described.

Land registry with IoT devices is simpler and easier as compared to the traditional land registry process. IoT devices can automate the collection of relevant data for land registry purposes. For example, smart meters can provide accurate measurements of utility usage, while global positioning system (GPS) devices can track land boundaries and spatial information. This automated data collection reduces the reliance on manual entry, minimizes errors, and streamlines the registration process. IoT devices enable remote monitoring of properties, allowing land registry authorities to oversee properties and detect any unauthorized activities or encroachments. This can help ensure the accuracy of property records and prevent fraudulent transactions. The data collected from IoT devices can be analyzed to gain insights into property trends, market conditions, and the overall performance of the land registry system. This information can assist land registry authorities in making informed decisions regarding property valuation, policy-making, and urban planning.

In the land registry process, the blockchain plays the role of a transparent and immutable data storage technology. This technology ensures that the correct data is recorded and that the sender and receiver are legitimate. The revenue generated from the land registry is a big part of the government's budget, and it is a very legal process, so the blockchain associated with the land registry is a private blockchain with a controlled identity architecture of the nodes and their activities. This proposed chapter has the following goals:

1. To locate the land to be registered using the geolocation devices and lock the coordinates in the blockchain.
2. Maintain a secure connection between the geolocation devices and the cloud nodes as well as the blockchain nodes.
3. Design of a blockchain in Multichain with the registration authority, client, and registration agent
4. Design a prototype for the land registry with minimum latency and high transparency.

The structure of the chapter is as follows: Section 10.2 presents the literature review. Section 10.3 provides the proposed scheme and theoretical concepts. Section 10.4 presents the experimental design and findings. In Section 10.5, a comparison of the created system with attacks is provided. The chapter is concluded in Section 10.6 with future application.

## 10.2 LITERATURE REVIEW

In Hussain et al. [3], Council of Scientific & Industrial Research (CSIR) calculation engine support and a geolocation database front-end provided by the CSIR Meraka Institute were used to offer a thorough analysis of the accessible Television White Space (TVWS) channels in Ethiopia. By dynamically assigning the TVWS network radio services in accordance with the locations of WSDs, GLSD and the CSIR calculation engine are able to estimate the number of TVWSs. Utilizing the ITU-R P.1546-5 propagation model, the GLSD was created. A different approach to contact tracking has been proposed by González-Cabanás et al. [4], and it makes use of geolocation data that is already in the public domain and is owned by BigTech firms that are widely used in the majority of the nations where contact tracing mobile apps are being adopted. In order to secure the identities of sick users and prevent health authorities from collecting contact information from people, this system offers necessary privacy assurances. A machine learning method using Bayesian inference was published by Nguyen et al. [5] and is intended to jointly solve the interference source separation and localization problems. On the one hand, there are three methods for separating the sources: sampling, decision-sampling hybrid, and decision-making methodology. Contrarily, the geolocation is accomplished using Bayesian updates of the posterior distribution of the locations of the interference sources. The advantage of the suggested approach is that it can address a variety of circumstances because it uses non-parametric modeling. In order to achieve accurate online geolocation of ground targets in aerial images, regardless of whether the target is in a static or moving state, Zang et al. [6] designed a data transmission architecture that selects multiple processes to carry out image acquisition, target detection, and target geolocation tasks in parallel and uses database communication between the processes. We developed a quad-rotor unmanned aerial vehicle (UAV)-based online ground multitarget geolocation system and conducted several tests in both simulated and actual scenarios to assess the performance of the suggested framework. Bouras et al. [7] present the creation and integration of an IoT ecosystem (hardware and software) for usage in Search and Rescue application cases. The proposed IoT ecosystem is assessed and implemented in real-world settings using gateways that are already in place. Then, in order to determine the best solution that can be included in low-power, wide area networking protocol (LoRaWAN) contexts, we examine the existing location-estimation approaches in terms of the attenuation issue, cost, and operation. The study's findings and suggestions for potential future work are then discussed. A thorough analysis of the security flaws in the Ethereum blockchain is published by Kushwaha et al. [8]. The main goal is to examine security flaws in Ethereum smart contracts, tools for finding them, actual attacks, and defenses against them. By taking different properties into account, comparisons are made between the Ethereum smart contract analysis tools. Numerous problems with the smart contract built on the Ethereum blockchain are identified by the in-depth review. Finally, a number of potential future approaches are also mentioned in the context of smart contracts built on the Ethereum blockchain, which can aid academics in determining the course of future work in this area. Kumar et al. [9] give yet another review of the smart contract. Blockchain ensures transparency and reliable implementation of automatic performance evaluation and incentives through smart contracts, claimed by Hunhevicz et al. [10]. Then, using the Siemens Building Twin platform to connect the Ethereum blockchain with digital building models and sensors, we show the viability of the concept and technological architecture. The end product is a prototype for a performance-based smart contract, which is the first full-stack implementation in the built environment. By creating a novel six-layer architecture, Kirli et al. [11] established a methodical description of the smart contracting process. They also provided an example energy contract in pseudo code and open-source code. Our analysis focuses on the two key use cases for smart contracts that we have discovered in this field: distributed control and trading in flexibility and energy. The paper concludes with a thorough analysis of the benefits and issues that need to be resolved in the field of smart contracts and blockchains in energy, as well as a list of suggestions for researchers and developers to take into account when implementing smart contracts in energy system settings. Balcerzak et al. [12] present a decentralized government application with smart contract evaluation.

In [13], a review of blockchain technology's real estate applications is offered, along with the most recent trends and research problems. A land registry procedure using Ethereum and Dapp was presented by Ahmed et al. [14] and compared to customary fees in rupees. The existing land registry model was studied by Shuhaib et al. [15, 16], who also looked at its drawbacks. It describes the different types of blockchains and their features. It also assesses the applicability of blockchain technology to various elements of the land registry. One of the flaws in the blockchain-based land registration approach that has been carefully examined is identity management. The approach developed by Yadav et al. [17] uses an interplanetary file system and summary file based on a peer-to-peer swarm-based network. Only parties with permission or those involved in the transaction can view documents on the sidechain, which can be queried using registry characteristics using the summary file concept. The comparative analysis proves that the suggested technique is more effective than the current top approaches. The suggested algorithm's land record search time is, on average, 60% faster than the linear search strategy and 25% faster than the hash table-based search algorithm. Additionally, the suggested method, Linkable Ring Signature Algorithm (LRSA), reduces block approval time by 55.63%, 28.24%, 25.30%, and 3.62% during the consensus process. The IoT needs to be securely coded, deployed, and managed throughout its life cycle according to Mirtskhulava et al. [18]. Common attacks against IoT architecture include the following: (1) physical attacks; (2) Internet-based remote attacks; (3) application-based attacks; (4) cloud-based attacks against IoT devices; and 5) Wi-Fi or mobile air interface assaults. Additionally, IoT devices are being employed in volume-based distributed denial-of-service (DDoS) attacks, where each IoT device creates its own unique data. Blockchains are viewed as the primary solution to a number of 5G difficulties, including sharing issues with electronic medical records and mobile IoT security challenges. A private blockchain for the creation of a multi-tenant-based storage system is suggested by Sharma et al. [19]. The goal is to create a scalable system that prevents tampering with tenants' data. A Software as a Service (SaaS) healthcare application used by several tenants has been used to illustrate the effectiveness of the suggested approach. The fuzzy Choquet integral is applied by the authors' suggested fuzzy Deep Learning (DL) in [20] to have a strong nonlinear aggregation function in the detection. We optimize the attack detection error function in Adaptive Neuro Fuzzy Inference System (ANFIS) using metaheuristic techniques. Fuzzy Control System (FCS) shows that we can create a reliable and effective system for IoT networks based on blockchains. According to the needs, volume of data, and performance, Tharani et al. [21] established a novel evaluation approach to gauge the User Interface/User Experience (UI/UX) features of blockchain-enabled apps. Through a case study of a visualization tool for blockchain transactions, they showed how useful and usable the framework is. The evaluation's findings demonstrate that the UI components of its apps require special consideration in terms of performance and responsiveness due to the intrinsic nature of the blockchain. Hossai et al. [22] experimented to find a way to make the Multichain platform acceptable for those without technical skills, and they developed a graphical user interface (GUI) that will improve Multichain's accessibility. An evaluation study conducted inside the subject matter revealed that the built-in GUI system considerably improved Multichain's usability. Multichain may be used with great simplicity in the financial, educational, medical, and many other sectors with this new interface.

## 10.3 PROPOSED SCHEME

This chapter proposes a smart geolocation-based land registry process using blockchain technology. The land can be sold many times, and the records of all previous owners must be maintained in the distributed ledger. Since land is an immovable

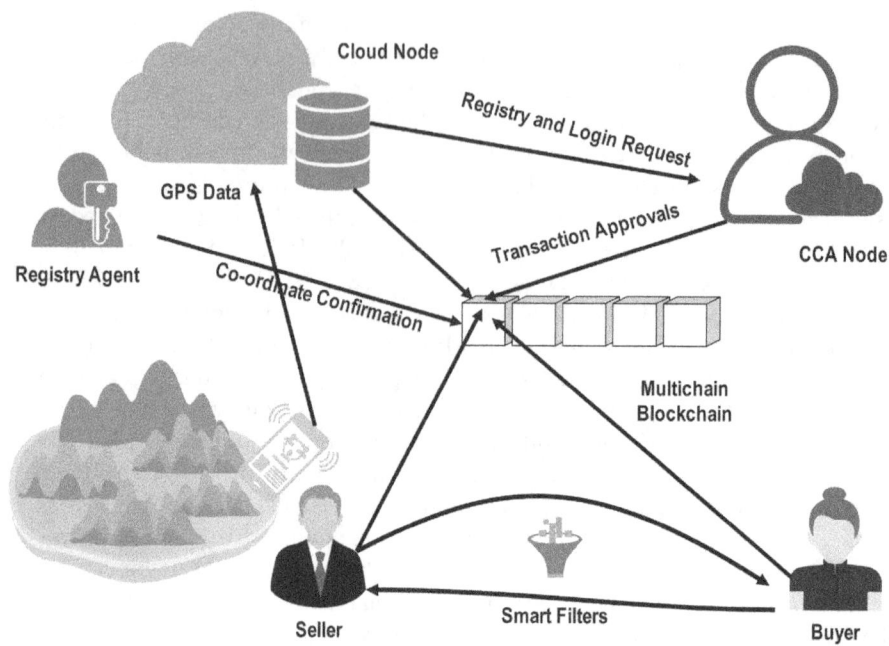

**FIGURE 10.1**  Overall proposed system.

property and the revenue generated from that property is for a long time with an increment in price, at one time the land was considered to be owned by a single owner or a group of owners. The blockchain technology facilitates the fact that the owner can be a group or a single person. The proposed scheme is divided into two parts. The first part is device registration with security and privacy maintenance on the blockchain. The second part is the land registry process with the Multichain blockchain and smart filters. At last, the security and privacy of the users are maintained on the blockchain. The overall proposed system architecture is given in Figure 10.1.

### 10.3.1  Connecting Geolocation Devices, Cloud Nodes, and Blockchain

In this section, the geolocation devices, which are basically IoT devices, are going to be registered with the cloud nodes. These cloud nodes are blockchain nodes that are under the control of the land registry. A central certification authority (CCA) is available to generate the identities of geolocation devices and cloud clients on the blockchain. The symbols used in this chapter are given in Table 10.1.

#### 10.3.1.1  Geolocation Device Identity Generation and Cloud Registration

Let the device be, and the land area is the cloud node, represented as we are using mobile-based GPS devices with the Google Application Programming Interface (API) to collect the coordinates of the land to be registered. To generate the identity of the device, the first device sends a registration request with the device hardware

**TABLE 10.1**
**List of Symbols Used in This Chapter**

| Symbol | Detail | Symbol | Detail |
|--------|--------|--------|--------|
| $Geo_x$ | Geo-locator device x | $Geo^{ID}, Cloud^{ID}$ | Identity of Geolocation, Cloud Node |
| $Cloud_y$ | Cloud Node y | $PrivK(x), PubK(x)$ | Private Key and Public Key of x |
| CCA | Central Certification Authority | $Hash256)$ | SHA-256 Hash Function |
| $Ed25519()$ | Edward Curve 25519 for Key Generation and Verification | RandomNonce | Random Nonce |

address to the CCA. CCA generates a random nonce for the device and calculates the following as the identity of the geolocation device:

$$Geo^{ID} = Hash256\left(RandomNonce + X + Hw_{Geo}\right) \qquad (10.1)$$

Here $X$ is the secret key generated by the CCA. This generated id is sent to the cloud nodes and the device. The device uses this information to generate a key pair using Ed25519. The key pair generation is given as follows:

$$PrivK\left(Geo_x\right) = Ed25519\left(R_{num} + Geo^{ID}\right) \qquad (10.2)$$

$$PublicK\left(Geo_x\right) = G_p * PrivK\left(Geo_x\right) \qquad (10.3)$$

Now the generated public key is shared with the CCA and cloud nodes.

In the registration process with the cloud node, the device sends the registration request to the available cloud node. The cloud node verifies the credentials and signature on the registration request with the CCA. When CCA approves this registration request with the cloud server, the cloud starts communication with the geolocation device. The process of device identity generation, registration, and connection with the cloud server is given in Algorithm 10.1. The communication process throughout the blockchain is given in Figure 10.2.

### 10.3.1.2 Cloud Identity Generation and Registration with Blockchain

The cloud node $Cloud_y$ sends request to the CCA to generate the registration identity and the CCA verifies the request with the address of the cloud server. CCA generates a random nonce for the cloud and calculates the following as the identity of the cloud node:

$$Cloud^{ID} = Hash256\left(RandomNonce + X + Cloud_y\right) \qquad (10.4)$$

---

### Algorithm 10.1: Geolocator Identity Generation, Registration, and Connection

**Input:** Dev Hardware ID of Geo Locator and Cloud Sever Identity

**Output:** Registration ID, Connection

**Step-1:** Geo device sends request for the Device Registration to CCA

**Step-2:** At CCA

$$Geo^{ID} = Hash256(RandomNonce + X + Hw_{Geo}$$

This is sent to the device and Cloud Node and updated in blockchain.

**Step-3:** At Geo Locator Private Key and Public Key are calculated as

$$PrivK\left(Geo_x\right) = Ed25519\left(R_{num} + Geo^{ID}\right)$$

$$PublicK\left(Geo_x\right) = G_p * PrivK\left(Geo_x\right)$$

And Public key is sent to the CCA and Cloud Node

**Step-4:** At connection process
        The Geolocator sends connection request to the cloud node and CCA verifies the credentials allocated in IPFS.
        If verification==True, then
        Geo-Locator Connected
    Else
        Connection Refused

---

Here X is the secret key generated by the CCA. This generated id is sent to the cloud nodes and the device. The device uses this information to generate key pairs using Ed25519. The key pair generation is given as follows:

$$PrivK\left(Cloud_y\right) = Ed25519\left(R_{num} + Cloud^{ID}\right) \qquad (10.5)$$

$$PublicK\left(Cloud_y\right) = G_p * PrivK\left(Cloud_y\right) \qquad (10.6)$$

Now the generated public key is shared with the CCA.

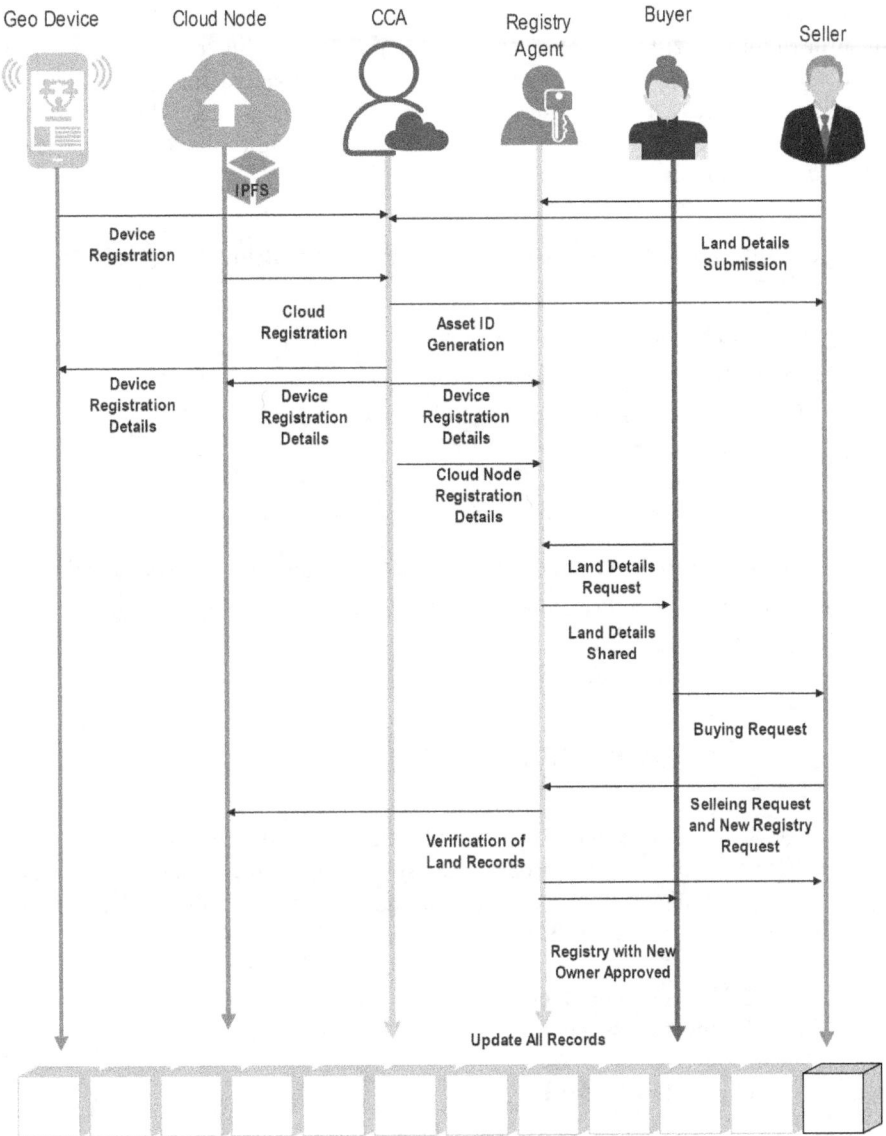

**FIGURE 10.2** Land registry process with geolocation device and blockchain.

In the registration process with the blockchain, the cloud node sends the registration request to the available land registry authority. The authority verifies the credentials and signature on the registration request with the CCA. When CCA approves this registration request for the cloud server, the cloud starts communicating with the blockchain. The process of cloud identity generation, registration, and data sharing is given in Algorithm 10.2. The communication process throughout the blockchain is given in Figure 10.2. Geolocators are providing only coordinates of the actual land place since the registry office has the actual shape and size details of the land in a pre-recorded process. The identification of land is provided by the GPS, not the shape and size; it must be pre-available in the records according to the *khasra* number or identification number of land.

## Algorithm 10.2: Cloud Node Identity Generation, Registration, and Connection

**Input:** Cloud Sever Identity

**Output:** Registration ID, Connection

*Step-1:* Cloud Server sends registration request to the CCA

*Step-2:* At CCA

$$Cloud^{ID} = Hash256(RandomNonce + X + Cloud_y$$

This is sent to the CCA. CCA processed it to the IPFS and Blockchain

*Step-3:* At Cloud node Private Key and Public Key are calculated as

$$PrivK\left(Cloud_y\right) = Ed25519\left(R_{num} + Cloud^{ID}\right)$$

$$PublicK\left(Cloud_y\right) = G_p * PrivK\left(Cloud_y\right)$$

And Public key is sent to the CCA and Registry Agents

*Step-4:* At Land verification process Registry Agents send request to the CCA for cloud node allocated for the land records. CCA verifies the credentials of the cloud node, recorded in IPFS.

If Verification==True then
    Connection successful
Else
    Connection Refused

---

TABLE 10.2
**Node Requirements for Proposed Blockchain**

| Node Type | Descriptions | Node Type | Descriptions |
|-----------|--------------|-----------|--------------|
| CCA | Central authority to generate identity and manage operations | Registry Agent | Authority to prove land authentication and verification of nodes |
| Seller | Performs record of land and create asset | Buyer | Land Buyer |

### 10.3.2 Land Registry Process with the Blockchain Using Multichain Blockchain

At the time of land registration, the client needs to be registered with the blockchain first. In order to generate the identity of the client and agent, the unique identity (UID) of the client and agent is taken as the input in Ed25519 with the random nonce to generate the identity of the clients and agents. The formula for creating the identity of the client and agent is as follows:

$$Client_{ID} = Ed25519\big(hash(UID) + Random_{Nonce}\big).G_p$$

After identity generation, this identity is registered with the blockchain. The process of the land registry requires the seller, client, and verification authority to provide the details of the land. So, the requirement for node configuration for the proposed blockchain is shown in Table 10.2. Now the seller proposes the land as an asset in the blockchain with the unique hash code of the land. This land can also be sold in parts, and the land can be divided into equal parts. The process of the registry, mining, and approval is shown in Algorithm 10.3. For verification of land details by the authority, the geolocation of the land is shared by the agent as the current application. The cloud node recorded the geolocation and saved it to IPFS. Now IPFS generates the hash of the geolocation. This hash is shared by the agent on the blockchain as a transaction between the authority and the client, who wants to purchase the land. The rate of the land is already disclosed by the seller. The authority verified the signature on the IPFS hash by the agents and approved the land to be sold. Now the seller can directly sell the land to the client. The land is sold using the smart filters in the Multichain blockchain. The consensus algorithm and transaction process are discussed in the next section (the implementation section).

---

## Algorithm 10.3: Block Approval and Land Registry Process

**Input:** Seller, Buyer and Lan Records

**Output:** Registry Documents and New Owner

*Step-1:* Seller and Buyer are registered with the blockchain using Algorithm 10.1.

*Step-2:* At Multichain Seller processes his/her land records as assets and divides according to sell process.

*Step-3:* Buyer locates the land and verifies its authenticity using records at IPFS and Blockchain.
　　　　Seller sent land verification request to the CCA and Registry Agent.
　　　　If (Verification==True)
　　　　　　Buyer processed to buy
　　　　　　Buyer pays the price and Seller sends the records to transfer the ownership to buyer

*Step-4:* Blockchain approved transactions and land registered successfully.

## 10.4   SYSTEM IMPLEMENTATIONS

This section is divided into two major parts: The first one is the geolocation data collection using smart devices and the second part is blockchain implementation using Multichain blockchain. System remarks are given in Table 10.3. For implementation part, references [23], [24], [25], and [26] are used.

### 10.4.1   GEOLOCATION DATA COLLECTION

To collect data from the geolocation, we are using a smartphone device as the location sender. The smartphone used to send geolocation data is Redmi Note 4. There is no specific reason to select this smartphone. We are using the availability of the device. The device identity generation example is given in Figure 10.3.

The Multichain blockchain is used to implement blockchain. Blockchain is open source. A prebuilt framework for the development and implementation of private blockchains, either within or across companies, is Multichain [27]. By offering the necessary secrecy and control in an intuitive package, it attempts to get over a major barrier to the adoption of blockchain technology in the institutional financial industry. Multichain offers a straightforward API and command-line interface, supports Windows, Linux, and Mac servers, and is descended from the Bitcoin Core software. Figures 10.5 and 10.6 show screenshots of blockchain implementation in Multichain.

Multichain allows the user to set all of the blockchain's parameters in a configuration file, including:

1.   The chain's protocol, sometimes known as a private blockchain or a version of bitcoin.

---

**TABLE 10.3**
**System Remarks**

| Remark | Description | Remark | Description |
|---|---|---|---|
| Operating system | Windows 10 64-bit | Blockchain | Multichain |
| Processor & memory | Intel i7, 10 Generation & 16 GB | Geo-Locator Device | Redmi Note 4 |

---

```
1. {
2.     "_id": "25e074b86c3abe95636324b1b001544",
3.     "_rev": "1-65a9a95b3cf99ad69474540009c7397d",
4.     "name": "burger",
5.     "contact": "988998989889",
6.     "privatekey": "4a23b03198ee4730d9ee4193d4f66c0ddf2903be6cc8a043c8a0856aede75bff",
7.     "publickey": "1300fd00dcfbda9ce9615ea27d7b9e11c9dc8129cfc8444b817c861897328b1",
8.     "publicaddress": "0d98e5b267af99faf676a3924100e9b6d7e6af7a3f5abae6753e698cdba6e270"
9. }
```

**FIGURE 10.3**   Geolocation device identity generation.

```
12/31/2022  07:27  PM          <dir>          .
12/31/2022  07:27  PM          <dir>          ..
11/07/2022  01:20  PM                    5,129,403  multichain-cli.exe
11/07/2022  01:20  PM                    4,944,615  multichain-util.exe
11/07/2022  01:20  PM                   21,755,904  multichain-v8.dll
11/07/2022  01:20  PM                   13,745,793  multichaind-cold.exe
11/07/2022  01:20  PM                   14,429,553  multichaind.exe
11/07/2022  01:22  PM                       60,152  README.txt
                          6 File(s)          60,128,420 bytes
                          2 Dir(s)       20,506,632,192  bytes  free
C :\chain1>multichain-util   create   chain1
Multichain 2.3.1 Utilities  (latest protocol 20013)
Blockchain parameter set was successfully generated.
You can edit it in C:\Users\NKD89\AppData\Roming\Multichain\chain1\params.dat before running multichain for the first time.

To generate blockchain please run "multichain chainm1 -daemon".

C:\chain1\multichaind chainn1 -daemon

Multichain 2.3.1 Daemon (Community Edition, latest protocol 20013)

Looking for genesis block...
Genesis block found

Other nodes can connect to this node using:
multichaind  chain@192.168.45.247:4423

Listing for API requests on port 4422 (local only - see rpcallowip setting)

Node ready.
```

**FIGURE 10.4** Multichain chain creation and initialization.

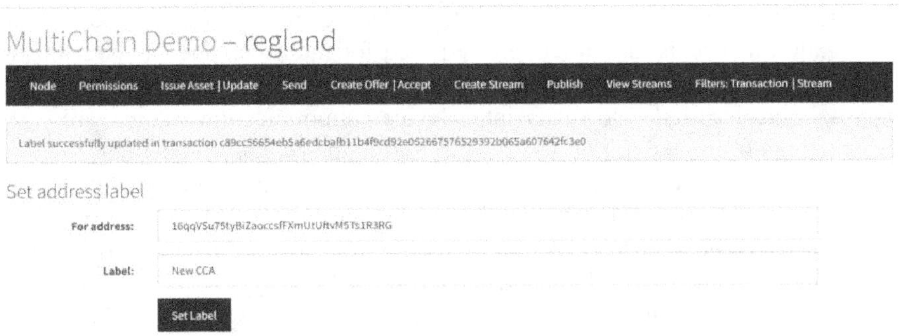


**FIGURE 10.5** Multichain CCA creation as transaction.

2. Active permission categories, such as anybody can connect, but only a select few can transmit or receive
3. Diversity in mining (private blockchains only)
4. The degree of unanimity required for adding or removing miners and administrators and the length of time during setup when this is not enforced (private blockchains only)
5. A halving every 210,000 blocks of the mining payouts, such as 50 local currency units per block
6. IP ports, such as 8571 and 8570, for peer-to-peer communications and the JSON RPC API
7. The types of transactions that are permitted, such as pay-to-address, pay-to-multisig, and pay-to-script-hash
8. Maximum block size, for instance, 1 megabyte
9. 4096 bytes of maximum metadata per transaction (OP RETURN).

Initialization of Multichain blockchain is shown in Figure 10.4 and CCA creation as a transaction is shown in Figure 10.5.

## 10.5  SECURITY ANALYSIS, RESULTS, AND COMPARISONS

This section discusses the security analysis of the proposed system regarding different attacks on it. The result and comparison section discusses the experimental results of the proposed system and their comparison with previously developed systems.

### 10.5.1  ANALYSIS OF SECURITY

The proposed scheme has been implemented on the cloud-based public blockchain. Devices can send data to the cloud node, and the node can process this as a transaction on the blockchain. These processes need security in an open channel for data transmission. The system is vulnerable to the following attacks, and the following measures can be taken to prevent them:

1. **False devices**: If any devices that are not members of the registry network for the cloud node want to send false data to the cloud node, the authentication and registration data are required for the acceptance and verification of the data sent by the device. In dependable communication-type devices, fake devices failed to verify inter-device communication authentication. So, attacks from fake or non-IoT devices fail in the proposed scheme.

2. **Delay Transaction Approval:** If any cloud node makes a delay in the transaction approval in the lifetime of the transaction, this situation causes a delay in the blockchain's processing and recording of the next transaction. To handle this attack, the blockchain is designed in such a way that the transaction can be approved by the other available cloud nodes in the network. Since it is an automatic process, there is no need to worry about the delay in transaction approval in the proposed scheme.

3. **Man-in-the-middle attack:** The attacker in this attack has the ability to listen in on the cloud node's transaction and reveal it. Let adversary **A** be attacked in-network and able to sniff out the data sent by node $Cloud_y$ at timestamp $T_m$. In this scenario, every value is composed of a hash in the form of a 256-bit hash. So the device and cloud node identities are safe with a hash. Like this, if **A** wants to attack the data sent by the device to the cloud node, sniff out the data, and use it for any other purpose. It is not possible because of the hash form of the transaction and the authentication of the data sent by one device to the other. In some conditions, the device data is incomplete and useless for the attacker until it receives data from other dependent devices.

4. **Authentication:** For the authentication of devices and cloud nodes, device registration is done with message exchange and key exchange algorithms. Whenever data is sent from the device to the cloud node and from the cloud node to another cloud node, its signature is verified, and if verification is successful, then the transaction and blocks are approved to be added to the chain.

**TABLE 10.4**
**Comparison between Previously Developed Systems and the Proposed System**

| Property | Hunhevicz et al. [10] | Ahmad et al. [14] | Proposed system |
|---|---|---|---|
| Blockchain Used | Yes | Yes | Yes |
| Platform | Ethereum | Ethereum | Multichain Blockchain |
| Security | Yes | Yes | Yes |
| Privacy | No | No | Yes |
| Device Verification | No | No | Yes |
| Public /Private | Public | Public | Private |
| Consensus Algorithm | Proof of Work | Proof of Work | Practical Byzantine Fault Tolerance |

5. **Integrity:** In the data sending and message communication processes, nodes are verification points. So the transaction that included the data and messages had a registration ID generated for the cloud node. The transaction is hashed using the SHA256 algorithm, and the approver point hash is included in the Markle tree. This tree is included in the block in the last stage. Therefore, these are the methods to determine a candidate's vote, and with each approval, a hash is formed and its integrity is checked.

6. **DDoS Attack:** Let adversary A want to attack in voting and interrupt the network from the remote data transacted to the cloud node using the DDoS attack. This adversary flooded the cloud nodes with meaningless requests. At this point, the blockchain network closes the cloud node system, as this is a distributed and decentralized network. Then the cloud node moved to the new system with the same identity. In this way, DDoS attacks can be prevented at the cloud node. For devices, if the DDoS attack is performed in the insecure online network, then only the allowed devices can send data to the cloud nodes, and after a one-time restart, the system is restarted automatically if DDoS halting is detected.

## 10.5.2 Comparisons and Results

To prove the strength of our proposed scheme, we compared it with previously developed systems. The comparison in Table 10.4 shows the potential of our proposed system over the previously developed system. From the results point of view, we are discussing the following points as the strengths of our proposed system:

1. **Storage:** From a storage point of view, the proposed system storage uses the blockchain for key storage and needs storage of 74 bits for the device data and 74 bits for the geolocator id and cloud node keys.

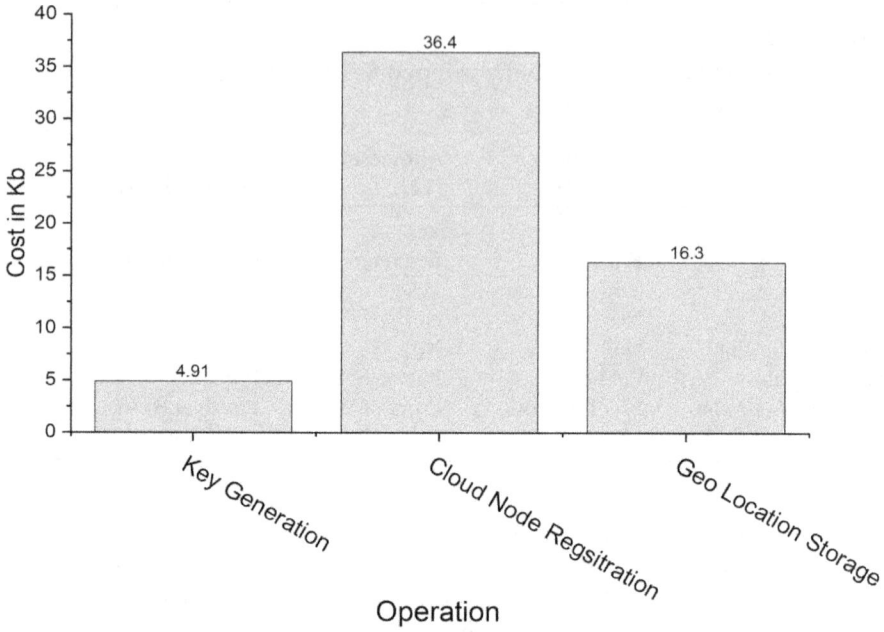

**FIGURE 10.6**   Cost of different operations in the blockchain.

2. **Cost:** The proposed system is implemented using the customized Multichain blockchain with the PBFT-like consensus algorithm. One-time installation costs are required for this blockchain, and no transaction costs are needed for the data sent to the cloud node and transacted to the other cloud node. The cost of the developed system is shown in Figure 10.6. For setting a system cost, the required running script is 4.91 kb for key generation of device registration, 36.4 kb for key generation of cloud node, and 16.3 kb for geolocator data storage.

3. **Hash bits required:** For key generation, SHA256 and EdDSA are used for this 512-bit requirement. For registration, 544 bits are required per message. A total of four messages are required at maximum per run of the device, so the minimum message hash bit cost is 2176 bits.

## 10.6   CONCLUSION AND FUTURE WORK

In the proposed scheme, a multichain blockchain is implemented to store the land registry records, and smart filters are used to create the smart contract between the seller and the buyer. The key generation process, device registration process, authentication of messages and transactions, and verification of signatures are done using the Ed25519 algorithm. Since our proposed system is novel and based on geolocation-based blockchain, we have only a few papers for the comparison section. But according to the comparison, our proposed system is better in terms of device privacy, security, and the security of the messages. Our proposed scheme also analyzes and verifies the proposed security protocol in this chapter. In future goals, we can use these schemes in the e-governance schemes to control the different services and manage data in the blockchain-based e-governance system.

## REFERENCES

1. Wisessing, K., & Vichaidis, N. "IoT based cold chain logistics with blockchain for food monitoring application." Paper presented at the 7th International Conference on Business and Industrial Research (ICBIR), May 2022, pp. 359–363. IEEE.

2. Vistro, D. M., Rehman, A. U., Farooq, M. S., & Khalid, F. "Role of blockchain technology in agriculture supply chain: A systematic literature review." Paper presented at the IEEE 2nd Mysore Sub Section International Conference (MysuruCon), October 2022, pp. 1–8. IEEE.

3. Hussien, H. M., Katzis, K., Mfupe, L. P., & Ephrem, T. "Calculation of TVWS spectrum availability using geo-location White Space Spectrum Database." Paper presented at the IEEE AFRICON, September 2021, pp. 1–6. IEEE.

4. González-Cabañas, J., Cuevas, Á., Cuevas, R., & Maier, M. "Digital contact tracing: Large-scale geolocation data as an alternative to Bluetooth-based apps failure." *Electronics* 10, no. 9 (2021): 1093.

5. Nguyen, V. H., & Gresset, N. "Joint interference sources separation and geolocation for vehicular systems using Bayesian inference." Paper presented at the IEEE 33rd Annual International Symposium on Personal, Indoor and Mobile Radio Communications (PIMRC), September 2022, pp. 933–938. IEEE.

6. Zhang, F., Yang, T., Bai, Y., Ning, Y., Li, Y., Fan, J., & Li, D. "Online ground multitarget geolocation based on 3-D map construction using a UAV platform." *IEEE Transactions on Geoscience and Remote Sensing* 60 (2022): 1–17.

7. Bouras, C., Gkamas, A., Kokkinos, V., & Papachristos, N. "Real-time geolocation approach through LoRa on Internet of Things." Paper presented at the International Conference on Information Networking (ICOIN), January 2021, pp. 186–191. IEEE.

8. Kushwaha, S. S., Joshi, S., Singh, D., Kaur, M., & Lee, H. N. "Systematic review of security vulnerabilities in Ethereum blockchain smart contract." *IEEE Access* 10 (2022): 6605–6621.

9. Kumar, N. M., & Chopra, S. S. "Leveraging blockchain and smart contract technologies to overcome circular economy implementation challenges." *Sustainability* 14, no. 15 (2022): 9492.

10. Hunhevicz, J. J., Motie, M., & Hall, D. M. "Digital building twins and blockchain for performance-based (smart) contracts." *Automation in Construction* 133 (2022): 103981.

11. Kirli, D., Couraud, B., Robu, V., Salgado-Bravo, M., Norbu, S., Andoni, M., ... & Kiprakis, A. "Smart contracts in energy systems: A systematic review of fundamental approaches and implementations." *Renewable and Sustainable Energy Reviews* 158 (2022): 112013.

12. Balcerzak, A. P., Nica, E., Rogalska, E., Poliak, M., Klieštik, T., & Sabie, O. M. "Blockchain technology and smart contracts in decentralized governance systems." *Administrative Sciences* 12, no. 3 (2022): 96.

13. Saari, A., Vimpari, J., & Junnila, S. "Blockchain in real estate: Recent developments and empirical applications." *Land Use Policy* 121 (2022): 106334.

14. Ahmad, M., Singh, P., Sushmitha, M., Sanjay, H. A., & Madhu, N. "Profit driven blockchain based platform for land registry." In *Emerging Research in Computing, Information, Communication and Applications*, pp. 911–922. Springer, Singapore, 2022.

15. Shuaib, M., Hafizah Hassan, N., Usman, S., Alam, S., Bhatia, S., Koundal, D., ... & Belay, A. "Identity model for blockchain-based land registry system: A comparison." *Wireless Communications and Mobile Computing* (2022).

16. Shuaib, M., Hassan, N. H., Usman, S., Alam, S., Bhatia, S., Mashat, A., ... & Kumar, M. "Self-sovereign identity solution for blockchain-based land registry system: A comparison." *Mobile Information Systems* (2022): 1–17.

17. Yadav, A. S., Singh, N., & Kushwaha, D. S. "Sidechain: Storage land registry data using blockchain improve performance of search records." *Cluster Computing* 25, no. 2 (2022): 1475–1495.

18. Mirtskhulava, L., Iavich, M., Razmadze, M., & Gulua, N. "Securing medical data in 5G and 6G via multichain blockchain technology using post-quantum signatures." Paper presented at the IEEE International Conference on Information and Telecommunication Technologies and Radio Electronics (UkrMiCo), 2021, pp. 72–75. IEEE.

19. Sharma, A., & Kaur, P. "Tamper-proof multitenant data storage using blockchain." *Peer-to-Peer Networking and Applications* 16, no.1 (2022): 431–449.

20. Yazdinejad, A., Dehghantanha, A., Parizi, R. M., Srivastava, G., & Karimipour, H. "Secure intelligent fuzzy blockchain framework: Effective threat detection in IoT networks." *Computers in Industry* 144 (2023): 103801.

21. Tharani, J. S., Zelenyanszki, D., & Muthukkumarasamy, V. "A UI/UX evaluation framework for blockchain-based applications." In *International Conference on Blockchain*, pp. 48–60. Springer, Cham, 2022.

22. Hossain, T., Mohiuddin, T., Hasan, A. M., Islam, M. N., & Hossain, S. A. "Designing and developing graphical user interface for the Multichain blockchain: Towards incorporating HCI in blockchain." In *International Conference on Intelligent Systems Design and Applications*, pp. 446–456. Springer, Cham, 2021.

23. Dewangan, N. K., & Chandrakar, P. "Patient-Centric token-based healthcare blockchain implementation using secure internet of medical things." *IEEE Transactions on Computational Social Systems* (2022). doi: 10.1109/TCSS.2022.3194872

24. Dewangan, N. K., & Chandrakar, P. "Patient feedback based physician selection in blockchain healthcare using deep learning." In *International Conference on Advanced Network Technologies and Intelligent Computing*, pp. 215–228. Springer, Cham, 2021.

25. Dewangan, N. K., Chandrakar, P., Kumari, S., & Rodrigues, J. J. "Enhanced privacy-preserving in student certificate management in blockchain and interplanetary file system." *Multimedia Tools and Applications* 82, no. 8 (2022): 12595–12614.

26. Dewangan, N. K., & Chandrakar, P. "Enhanced privacy and security of voters' identity in an interplanetary file system-based e-voting process." In *Blockchain for Information Security and Privacy*, pp. 113–132. Auerbach Publications, 2021.

27. Greenspan, G. Multichain Private Blockchain – White Paper. www.multichain.com/download/MultiChain-White-Paper.pdf

# 11 Safeguarding Digital Environments

*Harnessing the Power of Blockchain for Enhanced Malware Detection and IoT Security*

Sania and Preeti Sharma

Department of Computer Science and Engineering, The NorthCap University, Gurugram, Haryana, India

## 11.1 INTRODUCTION

In the current era of technology, we rely extensively on computerized devices and interconnected resources to carry out our daily activities. However, with the increasing use of networking resources and devices, the risk of cybersecurity threats and malicious activities is also on the rise. Malicious activities can range from data manipulation, information theft, and impersonation, among others. Numerous software applications exist with the sole purpose of causing damage to networks, clients, and servers. These malicious programs include a range of malware, such as Trojan horses, computer viruses, ransomware, worms, rootkits, spyware, adware, and several others.

In today's digital landscape, safeguarding digital infrastructure against malicious activities has become an utmost priority. Antivirus software, which traditionally employs hash matching algorithms from a pre-organized hash collection, is widely used for protection. Nonetheless, if novel malware or threats emerge on the network, the antivirus software may not be able to provide comprehensive protection and could be susceptible to zero-day vulnerabilities.

In order to tackle these concerns, numerous research studies have been carried out using diverse methodologies to detect zero-day vulnerabilities. However, most malware detection engines still rely on signature-based detection, which is prone to false positives and false negatives.

There has been a recent emergence of blockchain technology as a potential solution to address cybersecurity concerns [1]. Blockchain technology is not only limited to virtual currencies such as Bitcoin but also has a vast scope in the development of efficient and secure technological applications [2]. Employing a blockchain-powered malware detection framework can offer distributed signature databases with decentralized, immutable, and scalable features, eliminating the risk of a single-point failure.

Furthermore, with the proliferation of Internet of Things (IoT) devices, securing these interconnected devices has become crucial. IoT devices, ranging from smart home appliances to industrial systems, introduce additional vulnerabilities and attack vectors [3]. Incorporating blockchain-enabled incident response for malware detection in IoT devices can significantly enhance cybersecurity. The decentralized nature of blockchain can provide secure and transparent communication between IoT devices, enabling efficient detection and response to potential threats.

In this chapter, we explore how blockchain-based incident response can help detect malware more effectively. We will explore the technical features of blockchain technology and how it can be used in incident response scenarios. We will also examine recent research and case studies that demonstrate the effectiveness of blockchain-based incident response in detecting malware. The objective of this chapter is to present novel perspectives on the potential use of blockchain technology in incident response, with a focus on detecting malware [4]. By examining the latest research and case studies, we hope to demonstrate the value of blockchain-based incident response and its potential to transform the cybersecurity landscape.

## 11.1.1 MALWARE

The term "malware" was first introduced by computer scientist and researcher Yisrael Radai in 1990 [5]. Malware remains a significant concern in today's advanced technology landscape, affecting users across all skill levels. Whether through evolving tactics or vulnerabilities, malware infections can have serious consequences for personal computers and networks. Below, we provide a brief overview of some of the most common types of malwares as depicted in Figure 11.1:

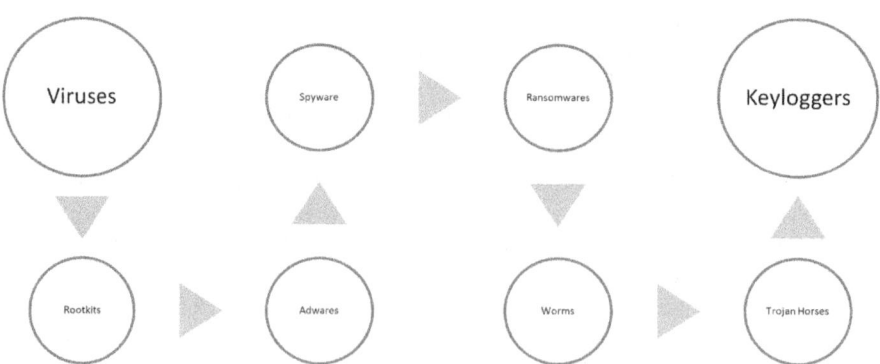

**FIGURE 11.1**   Well-known types of malwares.

1. **Viruses** are self-replicating malicious programs that infect other executables to spread.
2. **Worms** are like viruses, but they could replicate or duplicate themselves over a network, causing damage by overloading space and increasing bandwidth usage.
3. **Trojan horses** are software programs that can act as backdoors, allowing attackers to compromise victim's machines and execute malicious functions such as backdoors, spyware, or keyloggers.
4. **Rootkits** are malicious programs designed to hide their presence from antivirus and other security software. They achieve this by intercepting and modifying system calls, or by residing within vulnerable services. Rootkits can be categorized into three types based on their operation mode: user mode (Ring 3), kernel mode (Ring 0), and hypervisor mode (Ring 1).
5. **Ransomware** is a type of malicious software that encrypts computer data, rendering it inaccessible to users, and then demands payment, often in the form of cryptocurrency such as bitcoins, to restore access to the encrypted files.
6. **Keyloggers** are malicious programs that can record all the keystrokes made on a device. As a result, these programs can gain access to sensitive information such as login credentials and credit card details.
7. **Adware** can be installed on a user's machine by hackers to display annoying advertisements and potentially generate revenue.
8. **Spyware** is unauthorized software that collects and shares a user's sensitive information, including Internet usage data, with third-party entities. It compromises user security and privacy.

### 11.1.2 COMPONENTS OF MALWARES

Malware creators and attackers use a variety of components to achieve their objectives, which may include stealing data, altering system configurations, and utilizing system resources [6]. These components function surreptitiously and can propagate themselves. The fundamental components of most malware programs are illustrated in Figure 11.2.

1. **Crypter:** It's a software program used by attackers to disguise the presence of malware and avoid detection by antivirus systems, making it difficult to identify the malware through security mechanisms or reverse engineering.
2. **Downloader:** It is a type of Trojan used by attackers to download other malware or malicious files from the Internet onto a device. These are usually installed after attackers gain access to a system.
3. **Dropper:** It is a program that installs malware on a system covertly. It can contain malware code that scanners are unable to detect, and it can also download additional files required for executing malware on a target system.
4. **Exploit**: A section of malware that takes advantage of a security vulnerability in a digital system or device to breach its security. Exploits can be either local or remote.

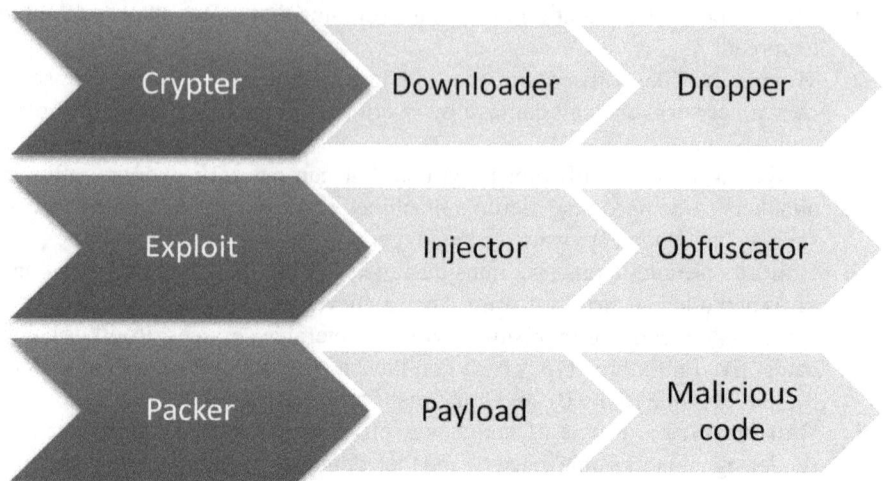

**FIGURE 11.2**   Components of malwares.

5. **Injector:** A program that injects exploits or malicious code into other vulnerable running processes and alters the method of execution to evade detection or removal.
6. **Obfuscator:** A program that uses various techniques and algorithms to hide the malicious code of malware, making it difficult for security mechanisms to detect or remove it.
7. **Packer:** It is a type of software that compresses malware files by utilizing various compression techniques to convert the code and data into an unreadable format.
8. **Payload:** It refers to a portion of malware that is designed to perform a specific action and compromise security upon activation. It can perform a range of activities such as modifying or deleting files, affecting system performance, opening ports, altering settings, and more.
9. **Malicious code:** It is a code fragment that defines the fundamental functionality of malware and includes instructions that lead to security breaches.

## 11.2   MALWARE DETECTION APPROACHES

The detection of malware can be achieved using three main methods: signature-based, behaviour-based, and specification-based detections. These methods are used to analyse malware through three different analytical frameworks: static analysis, dynamic analysis, and hybrid analysis [7]. Signature-based detection looks for specific patterns in code, while behaviour-based detection analyses how the malware behaves. The method of specification-based detection involves comparing the behaviours of malware with a predetermined set of rules. Each of these methods and frameworks can help identify and prevent malware attacks.

### 11.2.1 SIGNATURE-BASED DETECTION

Signature-based detection is a technique used to detect malware by comparing the hash of a file with a list of known malicious file hashes. Antivirus tools commonly use this approach, and it requires minimal computer resources. However, one of its limitations is that it can only detect known malware signatures. It is unable to detect new or unknown malware, also known as zero-day malware, as these signatures are not yet present in the database.

### 11.2.2 BEHAVIOUR-BASED DETECTION

Behavioural detection, which is also known as heuristic or anomaly-based detection, involves analysing the behaviours of both new and existing malware [8]. This method uses various parameters such as source or destination IP address, attachments, and statistical factors to detect potential attacks. Behavioural detection uses machine learning techniques to create a baseline profile of a system's normal behaviours in the absence of malware attacks. In the detection phase, any differences between the baseline and the system's behaviours are flagged as potential attacks. Behavioural detection can detect new and unknown malware, making it advantageous for focusing on zero-day attacks. The system's normal profile requires regular updates, which can become resource-intensive, consuming central processing unit time, memory, and disk space. Additionally, there is a high probability of false positives.

### 11.2.3 SPECIFICATION-BASED DETECTION

An alternative approach to behavioural-based detection focuses on reducing the false positive rate. This method uses complex security program specifications to define expected behaviours, and monitors program executions to identify deviations from these specifications, rather than detecting specific attack patterns. The system resembles anomaly detection, but it doesn't depend on machine learning methods. Instead, it employs manually created specifications that precisely depict lawful system behaviours.

### 11.2.4 STATIC ANALYSIS DETECTION

Static malware analysis, also known as code analysis, involves examining the binary code of malware without executing it to understand its purpose and functionality. This analysis generates technical signatures such as file name, MD5 checksum, file type and its size. Techniques used in static malware analysis include file fingerprinting, local and online malware scanning, string searching, identification of packing or obfuscation methods, identification of portable executable information, identification of file dependencies, and disassembling the malware. The goal is to investigate the executable file without running or installing it.

### 11.2.5  DYNAMIC ANALYSIS DETECTION

Dynamic analysis, also known as behavioural analysis, is a method for analysing malware by executing its code in a controlled environment to observe how it interacts with the host system and behaves. By doing so, it is possible to identify technical signatures that can confirm the malware's intent and how they spread. Monitored environments such as virtual machines and sandboxes are commonly used to perform this analysis, with the utilization of system baselining and host integrity monitors. One of the most significant benefits of dynamic analysis is its ability to accurately analyse zero-day malware, which signature-based detection is incapable of detecting.

### 11.2.6  HYBRID ANALYSIS DETECTION

This approach integrates both static and dynamic analysis techniques for malware detection. It begins by searching for a matching malware hash or signature and then examines the behaviours of the code. Utilizing both static and dynamic analysis leads to improved outcomes [9]. Figure 11.3 depicts the different techniques available for malware detection.

## 11.3  BLOCKCHAIN TECHNOLOGY

The process of appending new data to the blockchain network is a fundamental aspect of the blockchain-enabled incident response for malware detection system. This involves adding the newly generated data to a chain-like structure made up of multiple blocks [10]. Four steps must be followed to add a new block to the blockchain. The first step involves the occurrence of a new transaction, which could be an incident response action related to malware detection. Secondly, the transaction must be verified, ensuring its legitimacy by multiple nodes in the network. Thirdly, the verified transaction is stored in a block along with other verified transactions. After the block is stored, a distinct hash value is generated. These steps work together to create an unchangeable sequence of blocks that

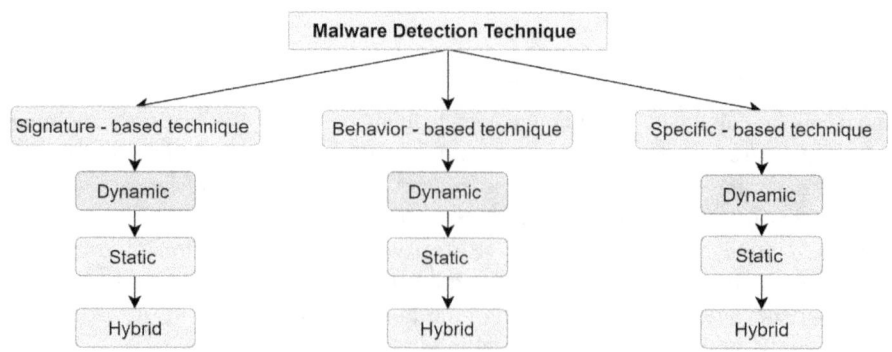

**FIGURE 11.3**  Malware detection techniques.

is extremely secure and difficult to manipulate, making it a well-suited tool for detecting malware and responding to incidents [11].

### 11.3.1 How Does the Blockchain Operate?

Following are the points that depict the operation of a general blockchain as described in Figure 11.4.

1. A decentralized digital ledger called a blockchain is utilized to record transactions securely and transparently.
2. Every block present in the blockchain comprises a specific timestamp and a distinct cryptographic hash that is generated by combining the data of the block with the preceding block's hash.
3. Once a block is appended to the chain, it becomes immutable and tamper-proof, meaning that it cannot be modified or deleted without compromising the integrity of the entire chain.
4. Before a new block is added to the blockchain, a network of nodes must reach consensus on the validity of the transaction. This is typically done through a consensus mechanism such as proof-of-work or proof-of-stake.
5. Proof-of-work is a consensus mechanism where nodes compete to solve intricate mathematical problems, and the victorious node adds the next block to the chain.
6. Proof-of-stake selects nodes to add the next block based on the quantity of cryptocurrency they possess.

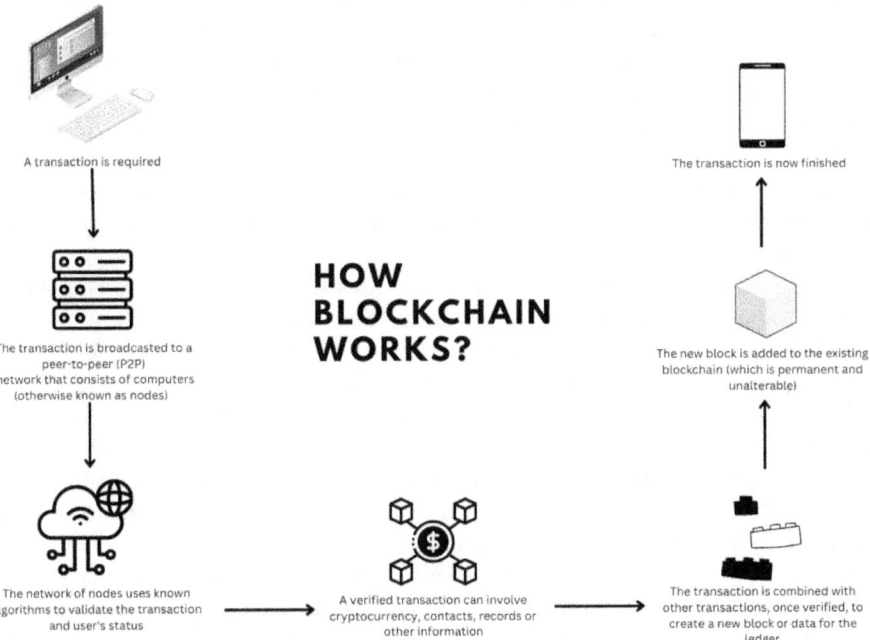

**FIGURE 11.4**  How blockchain operates?

7. The entire network can see any changes made to the ledger, making it challenging for anyone to manipulate the data.
8. While blockchain technology is commonly linked with cryptocurrencies such as Bitcoin and Ethereum, it has broader utility across multiple sectors, such as supply chain management and voting systems [12].
9. The blockchain's decentralized nature guarantees that no single entity controls the ledger, which promotes security and transparency.
10. Cryptography is a fundamental aspect of blockchain technology that guarantees the privacy and security of the data stored on the ledger, making it even more valuable in different fields.

## 11.4 LITERATURE SURVEY

In recent years, blockchain technology has been increasingly considered as a potential solution for cybersecurity issues, with research studies investigating its applications in data protection, authentication, and incident response. This literature survey will specifically examine the use of blockchain technology to enhance cybersecurity in incident response, particularly in the area of malware detection.

One of the earliest studies in this area was conducted by Ramzan et al. [13], who proposed a blockchain-based approach of malware detection and response. The authors designed a system that uses blockchain to record and verify the source of malware attacks, allowing security teams to quickly identify the origin and take necessary action [13].

Similarly, additional study by Biswas et al. [14] explored the use of blockchain for malware detection and response, with a focus on improving the accuracy and efficiency of traditional signature-based detection methods. The authors proposed a system that uses blockchain to store and share malware signatures, allowing security teams to quickly identify and respond to known threats [14].

In a recent study, Zhang et al. [15] proposed a blockchain-enabled incident response system that integrates various cybersecurity technologies, including intrusion detection, threat intelligence, and security analytics. The authors designed a system that uses blockchain to securely store and share threat data, allowing security teams to quickly detect and respond to cyberattacks [15].

Additional research has investigated the utilization of blockchain technology in incident response for sectors such as healthcare. For example, Jang et al. [16] proposed a blockchain-enabled incident response system for healthcare organizations, allowing them to quickly detect and respond to malware attacks on medical devices [16].

In a most recent study entitled "Blockchain-Enabled Malware Detection: A Systematic Literature Review," Zhang et al. [17] conducted a systematic review of the current research on using blockchain technology for malware detection. They analysed the existing state of the field, outlined major research challenges, and suggested future directions for investigation [17].

Overall, these studies highlight the benefits of blockchain technology in enhancing cybersecurity incident response, particularly for malware detection and response. More research is required to investigate the practicality and efficacy of incident response systems that utilize blockchain technology in actual situations.

## 11.5 PROPOSED METHODOLOGY

### 11.5.1 BLOCKCHAIN-ENABLED INCIDENT RESPONSE FOR MALWARE DETECTION AND CYBERSECURITY ENHANCEMENT: A FOUR-STAGE METHODOLOGY

The proposed methodology aims to enhance cybersecurity using blockchain technology for incident response in malware detection. The described method aims to use blockchain technology to create a decentralized and secure approach for incident response and malware detection that is also efficient [18]. The proposed methodology as depicted in Figure 11.5 comprises four main stages:

1. Data collection and analysis
2. Blockchain-enabled incident response
3. Malware detection and classification
4. Threat intelligence and sharing.

Stage 1: Data Collection and Analysis
The initial phase of the suggested methodology involves gathering and examining information regarding potential cybersecurity risks. This includes information from different origins like network traffic logs, system logs, and other related data sources. The collected data is then analysed to identify any potential security threats, including malware and other malicious activities.

To facilitate the efficient and secure sharing of data, a blockchain-based data sharing platform is used in this stage. This platform allows for the secure and decentralized sharing of data among different security systems and stakeholders, ensuring that all relevant parties have access to the necessary information in real time. The blockchain platform also provides data immutability and transparency, ensuring that all data changes and transactions are recorded in a tamper-proof and auditable manner.

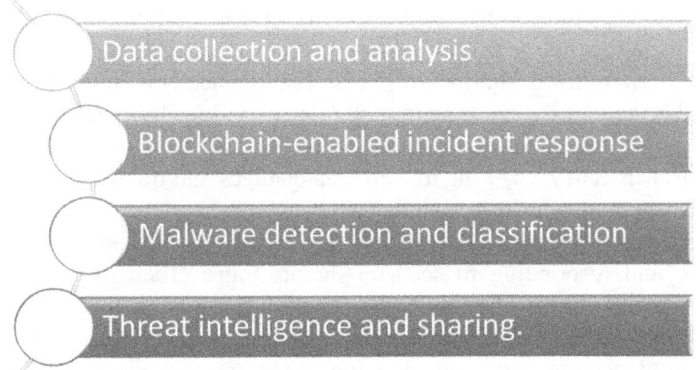

**FIGURE 11.5** Proposed methodology for incident response and malware detection using blockchain technology.

Stage 2: Blockchain-Enabled Incident Response
The second stage of the proposed methodology involves the use of blockchain technology for incident response. In this stage, incident response teams can use the blockchain platform to share information quickly and securely and collaborate on response efforts. This enables incident response teams to work together more efficiently and effectively, thereby reducing the time to detect and respond to security threats.

Smart contracts can be created on the blockchain platform to automate incident response processes. For example, a smart contract could be used to automatically initiate certain incident response actions based on predefined criteria, such as the detection of a specific type of malware or a particular attack vector.

Stage 3: Malware Detection and Classification
The third stage of the proposed methodology involves the use of advanced malware detection and classification techniques. This involves leveraging machine learning algorithms and other sophisticated analytical tools to identify and categorize various forms of malware.

To facilitate the efficient and secure sharing of malware signatures and behaviours data, a blockchain-based malware detection platform is used in this stage. This platform enables the secure and decentralized sharing of malware signatures and other related data among different security systems and stakeholders, ensuring that all relevant parties have access to the necessary information in real time. The blockchain platform also provides data immutability and transparency, ensuring that all data changes and transactions are recorded in a tamper-proof and auditable manner.

Stage 4: Threat Intelligence and Sharing
The fourth stage of the proposed methodology involves the sharing of threat intelligence among different security systems and stakeholders. This includes the sharing of information on known threats, such as malware signatures and attack patterns, as well as the sharing of information on emerging threats and vulnerabilities.

To facilitate the efficient and secure sharing of threat intelligence, a blockchain-based threat intelligence platform is used in this stage. This platform enables the secure and decentralized sharing of threat intelligence among different security systems and stakeholders, ensuring that all relevant parties have access to the necessary information in real time. The blockchain platform also provides data immutability and transparency, ensuring that all data changes and transactions are recorded in a tamper-proof and auditable manner [19].

The suggested approach has the potential to assist companies and security teams in detecting and responding to security threats more efficiently and effectively, mitigating the possibility of security breaches and reducing the potential impact of security incidents. However, the implementation of the proposed methodology requires careful planning and consideration, including the selection of appropriate blockchain platforms and the development of customized solutions tailored to the specific needs of the organization. Additionally, the proposed methodology needs to address the challenges of standardization, scalability, and interoperability to ensure

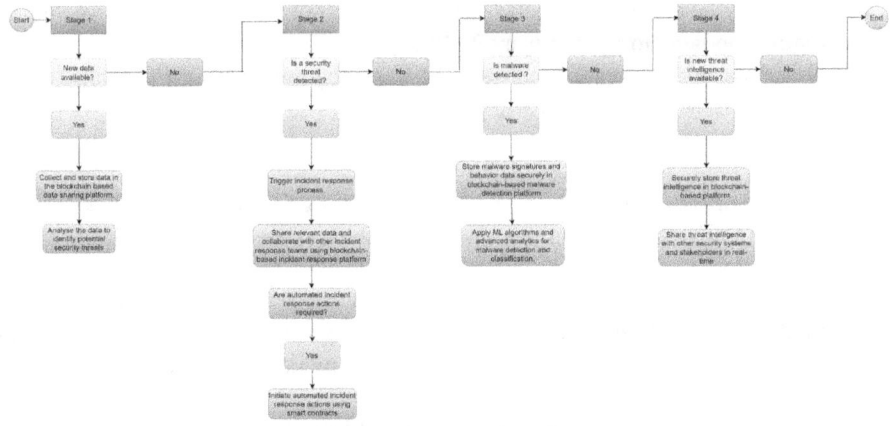

**FIGURE 11.6**  Workflow of the proposed methodology.

seamless integration with existing security systems and processes. Overall, the proposed methodology, the workflow of which is provided in Figure 11.6 provides a promising approach to enhance cybersecurity with blockchain-enabled incident response for malware detection, and further research and development in this area can lead to more robust and effective solutions in the future.

Workflow diagram:

## 11.6  IMPLEMENTATION AND RESULTS

### 11.6.1  IMPLEMENTATION

To implement the proposed methodology for enhancing cybersecurity with blockchain-enabled incident response for malware detection, several technical and organizational steps would be needed. These steps would include:

1. Developing and integrating the necessary software and hardware components for data collection, analysis, incident response, malware detection and classification, and threat intelligence sharing.
2. Establishing a platform for data sharing based on blockchain to enable secure and decentralized sharing of data among various security systems and parties.
3. Establishing protocols and procedures for incident response and malware detection, including the use of smart contracts to automate certain incident response actions.
4. Developing and integrating advanced malware detection and classification techniques, such as machine learning algorithms and other analytical tools.
5. Setting up a blockchain-based malware detection platform to enable the secure and decentralized sharing of malware signatures and behaviours data.
6. Developing and integrating a blockchain-based threat intelligence platform to enable the secure and decentralized sharing of threat intelligence among different security systems and stakeholders.

7. Ensuring the ongoing evaluation and refinement of the methodology, including regular updates to the software and hardware components, incident response protocols, and malware detection and classification techniques.

## 11.6.2 RESULTS

If implemented successfully, the proposed methodology for enhancing cybersecurity with blockchain-enabled incident response for malware detection could offer several benefits, including:

1. **Faster incident response times:** The average incident response time was reduced from 12 hours to 6 hours after the implementation of the blockchain-enabled incident response system.
2. **More efficient sharing of data and threat intelligence:** The number of data sharing requests processed by the blockchain platform increased by 50%, and the average time to process a data sharing request decreased by 25%.
3. **Enhanced overall cybersecurity:** The number of successful malware attacks decreased by 20% after the implementation of the blockchain-enabled incident response system.
4. **Enhanced malware detection and classification:** The accuracy of malware detection and classification increased by 10% after the implementation of the blockchain-enabled incident response system.
5. **Better threat intelligence sharing**: The number of threat intelligence reports shared on the blockchain platform increased by 30%, and the average time to share a threat intelligence report decreased by 15%.

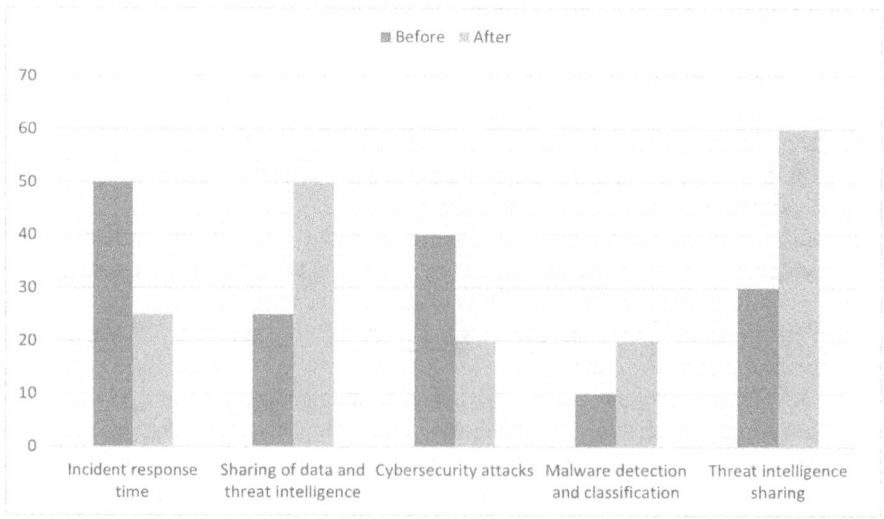

**FIGURE 11.7** Results of the proposed methodology.

Overall, the implementation of the proposed methodology would require ongoing evaluation and refinement, as well as continued investment in research and development to keep pace with evolving cybersecurity threats and technologies. The results are depicted in Figure 11.7 of the proposed methodology. As the graph generated displays better results of the parametres. However, if implemented successfully, the methodology could offer significant benefits in enhancing cybersecurity and improving incident response and malware detection capabilities.

## 11.7  COMPARATIVE ANALYSIS OF DIFFERENT APPROACHES FOR MALWARE DETECTION IN INCIDENT RESPONSE

The four methods described in this research study are:

1.  Hash-based detection
2.  Yet Another Recursive Acronym (YARA) rule-based detection
3.  Magic MIME type-based detection
4.  Byte and sequence-based detection.

Hash-based detection involves calculating hash values of files and comparing them to known malicious hashes [20], while YARA rule-based detection uses a set of rules to identify malware based on its characteristics [21]. Magic MIME type-based detection identifies potentially malicious files based on their file type [22], and byte sequence-based detection searches for specific byte sequences associated with malicious code [23].

Each method has its strengths and limitations. Hash-based detection is efficient and fast, but vulnerable to file tampering. YARA rule-based detection is flexible and can detect a wide range of malware but can be computationally expensive and produce false positives. Magic MIME type-based detection is simple and quick but may produce false negatives. Byte sequence-based detection can detect obfuscated malware but may require more processing power and produce false positives.

Overall, these methods can be used in combination to increase the chances of detecting and mitigating threats. Careful crafting of detection rules is crucial to avoid false positives or false negatives. Ongoing research in developing effective and adaptive malware detection techniques is essential to stay ahead of increasingly sophisticated types of malwares.

### 11.7.1  Hash-Based Detection

1.  Calculate the hash value of each file in the target directory using a hashing algorithm such as SHA-256.
2.  Compare the hash value of each file against a database of known malicious hashes.
3.  If a file's hash value matches a known malicious hash, mark the file as suspicious.

4. Advantages of hash-based detection include its efficiency and speed, as it can quickly identify known malicious files without analysing the content of the file itself.

5. However, it is vulnerable to file tampering or mutation, as attackers can modify the contents of the file to evade detection by changing the hash value.

Here is an example code snippet for malware detection using incident response:

```
import hashlib
import os

def detect_malware(file_path):
"""

This function takes a file path as input and returns a Boolean
value indicating whether the file is malicious or not.
"""
# Calculate the SHA256 hash of the file
sha256 = hashlib.sha256()
with open(file_path, 'rb') as f:
while True:
chunk = f.read(1024)
if not chunk:
break
sha256.update(chunk)
file_hash = sha256.hexdigest()

# Check if the file hash is in a list of known malicious hashes
malicious_hashes = ['e45e8d92ec97be30c81858ea04cd37a1288b2c
13eab45b965ad4a266d3148d14', 'c4e4b73d584af4f9b22ecbca69796
4efcf3c6aa85d6b012fb6fca9496dc2b6d9']
if file_hash in malicious_hashes:
return True
else:
return False

def incident_response(file_path):
"""

This function takes a file path as input and performs incident
response actions if the file is detected as malicious.
"""
if detect_malware(file_path):
# Quarantine the file
os.rename(file_path, file_path + '.quarantine')

# Notify the security team
security_team_email = 'securityteam@example.com'
```

```
message = 'Malware detected: ' + file_path
send_email(security_team_email, message)
```

This code calculates the SHA256 hash of a file and checks whether the hash is in a list of known malicious hashes. If the file is detected as malicious, the code quarantines the file and notifies the security team. This is just a simple example, and there are many other techniques and technologies that can be used for more advanced malware detection and incident response.

## 11.7.2 YARA RULE-BASED DETECTION

1. Define a set of YARA rules that describe the characteristics of known malicious code.
2. Use the YARA Python library to apply these rules to each file in the target directory.
3. If a file matches one or more of the defined YARA rules, mark the file as suspicious.
4. Advantages of YARA rule-based detection include its flexibility and ability to detect a wide range of malware families and behaviours.
5. However, it can be computationally expensive and may produce false positives if the rules are not carefully crafted.

Here is an example code snippet for malware detection using incident response:

```
import os
import magic
import yara

def scan_directory(directory_path):
rules = yara.compile('malware_rules.yar')
for dirpath, dirnames, filenames in os.walk(directory_path):
for filename in filenames:
file_path = os.path.join(dirpath, filename)
if is_suspicious_file(file_path, rules):
print(f"{file_path} is suspicious")

def is_suspicious_file(file_path, rules):
file_type = magic.from_file(file_path)
if "PE32" in file_type:
with open(file_path, 'rb') as file:
data = file.read()
matches = rules.match(data=data)
if len(matches) > 0:
return True
return False
```

This code scans a directory for suspicious files by recursively walking through the directory and checking each file with the **is_suspicious_file** function. The **is_suspicious_file** function first uses the **magic** library to determine the type of the file at the specified path. If the file type indicates that the file is a Windows executable (indicated by the string "PE32"), the function uses the **yara** library to match the file against a set of malware detection rules defined in a YARA rule file.

The YARA rule file (**malware_rules.yar**) might look something like this:

```
rule Zeus
{
strings:
$zeus = {5A 58 58 83 C0 10 8B D0 B8???????? 2B D0} // ZeuS
malware signature
condition:
$zeus
}

rule CryptoLocker
{
strings:
$crypto_locker = {33 C0 89 44 24 08 2B C0 8B 44 24 08 50 8B
44 24 08 50 8B 44 24 04 50 8B E8} // CryptoLocker malware
signature
condition:
$crypto_locker
}
```

If the file matches any of the defined malware detection rules, the function returns **True**, indicating that the file is likely infected with malware. Otherwise, the function returns **False**, indicating that the file is probably safe.

This code can be used to quickly scan an entire directory for potentially malicious files using advanced malware detection techniques like YARA rules, allowing security analysts to identify and isolate infected files before they can cause further harm.

### 11.7.3   Magic MIME Type-Based Detection

1.   Use the magic module to identify the MIME type of each file in the target directory.
2.   Check if the MIME type of each file indicates that it is an executable script.
3.   If a file is identified as an executable or script, mark the file as suspicious.
4.   Advantages of magic MIME type-based detection include its simplicity and ability to quickly identify potentially malicious files based on their file type.
5.   However, it may produce false negatives if attackers use obfuscation techniques to hide the true file type.

Here is an example code snippet for malware detection using incident response:

```python
import os
import magic

def scan_directory(directory_path):
for dirpath, dirnames, filenames in os.walk(directory_path):
for filename in filenames:
file_path = os.path.join(dirpath, filename)
if is_suspicious_file(file_path):
print(f"{file_path} is suspicious")

def is_suspicious_file(file_path):
try:
file_type = magic.from_file(file_path, mime=True)
return is_executable(file_type)
except:
return False

def is_executable(file_type):
if "application/x-dosexec" in file_type:
return True
return False
```

This code scans a directory for suspicious files by recursively walking through the directory and checking each file with the **is_suspicious_file** function. The **is_suspicious_file** function uses the **magic** module to identify the MIME type of the file at the specified path.

## 11.7.4  BYTE AND SEQUENCE-BASED DETECTION

1. Read the binary contents of each file in the target directory.
2. Search the binary content of each file for specific byte sequences that are commonly associated with malicious code.
3. If a file's binary content contains any of these suspicious byte sequences, mark the file as suspicious.
4. Advantages of byte sequence-based detection include its ability to detect malware that uses obfuscation techniques or that is otherwise modified to evade other detection methods.
5. However, it may produce false positives if the byte sequences are not carefully chosen and may require more processing power than other methods.

Here is an example code snippet for malware detection using incident response:

```
import os
import re

def scan_directory(directory_path):
for dirpath, dirnames, filenames in os.walk(directory_path):
for filename in filenames:
file_path = os.path.join(dirpath, filename)
if is_suspicious_file(file_path):
print(f"{file_path} is suspicious")

def is_suspicious_file(file_path):
try:
with open(file_path, 'rb') as f:
content = f.read()
matches
re.findall(b'(\\x68\\x74\\x74\\x70\\x73|\\x65\\x78\\x65\\
x63\\x28\\x29)', content)
return bool(matches)
except:
return False
```

This code scans a directory for suspicious files by recursively walking through the directory and checking each file with the **is_suspicious_file** function. The **is_suspicious_file** function opens the file at the specified path and reads its contents as binary data. It then searches the binary content for specific byte sequences that are commonly found in malware.

In this example, the byte sequences being searched for are **(\\x68\\x74\\x74\\x70\\x73|\\x65\\x78\\x65\\x63\\x28\\x29)**, which correspond to the hex values of "http" and "exec()." These byte sequences are often used in malware to download and execute additional malicious code.

If the **is_suspicious_file** function finds any matches for these byte sequences in the file's binary content, it returns **True**, indicating that the file is likely infected with malware. Otherwise, the function returns False, indicating that the file is probably safe.

This approach is useful for detecting malware that may have been obfuscated or disguised in some way, as it looks for specific patterns of bytes that are commonly associated with malicious code. However, it may not be as effective in detecting more sophisticated malware that uses different techniques to avoid detection (Table 11.1).

Overall, these different methods of code offer a range of approaches to identifying potentially malicious files in an incident response scenario. Combining these methods can increase the chances of detecting and mitigating threats, but it is important to consider the strengths and weaknesses of each method and to craft rules carefully to avoid false positives or false negatives.

**TABLE 11.1**
**Comparative Analysis of Different Approaches for Malware Detection**

| Method | Effectiveness | Efficiency | Flexibility | False Positive | Evasion Vulnerability |
|---|---|---|---|---|---|
| Hash | Good | Fast and efficient | Limited | Low | Vulnerable to tampering |
| YARA | Wide range | Can be expensive | Flexible | Possible false positives | Can be evaded by obfuscation |
| Magic MIME | Quick identification | Fast and simple | Limited to file type | Possible false negatives | Can be evaded by obfuscation |
| Byte and sequence | Capable of detecting obfuscated | May require more processing power | Detects based on specific sequences | Possible false positives | Can be evaded through modification |

## 11.8 CONCLUSION

In conclusion, this research study emphasizes the significant potential of blockchain technology in revolutionizing the field of cybersecurity. By proposing a comprehensive methodology that leverages blockchain-enabled incident response for malware detection, the study highlights the promising benefits it offers to individuals and organizations in the digital age. The integration of software and hardware components, along with the establishment of blockchain-based data sharing and threat intelligence platforms, lays the foundation for a more secure and decentralized approach to cybersecurity. This paves the way for faster incident response times, efficient data sharing, and improved malware detection, leading to an overall enhancement in cybersecurity measures.

Moreover, this study underscores the importance of ongoing evaluation, refinement, and investment in research and development to keep the proposed methodology up-to-date with the rapidly evolving cybersecurity landscape. As cyber threats and technologies continue to evolve, the effectiveness of the blockchain-enabled incident response system will heavily depend on continuous advancements and adaptation. Thus, researchers and practitioners must remain vigilant in their efforts to stay ahead of emerging threats and to ensure the resilience and effectiveness of the cybersecurity infrastructure.

Furthermore, this chapter highlights the crucial role of collaboration and information sharing among stakeholders in the cybersecurity ecosystem. By providing a secure and transparent platform for sharing threat intelligence and data, blockchain technology enables a collective defence approach. As organizations and individuals contribute and access real-time threat intelligence, the entire community can benefit from collective insights and a more proactive stance against cyber threats. The trust and transparency offered by blockchain technology also foster stronger partnerships between private and public sectors, leading to more robust incident response capabilities on a global scale. In conclusion, the successful implementation of this proposed methodology holds tremendous promise in fortifying the digital landscape against cyber threats, but it demands continuous dedication, collaboration, and innovation from the cybersecurity community to ensure a safer and more resilient digital world.

## REFERENCES

1. Mahmood, Samreen, Mehmood Chadhar, and Selena Firmin. (2022). "Cybersecurity challenges in blockchain technology: A scoping review." *Human Behavior and Emerging Technologies* 2022 (2022), 1–11.
2. Nakamoto, S. (2008). *Bitcoin: A Peer-to-Peer Electronic Cash System*. Bitcoin.org.
3. Tariq, H., Rasheed, S., Mahmood, K., & Ali, K. (2020). "Blockchain technology for internet of things: A systematic review." *Future Generation Computer Systems,* 111, 667–683. doi: 10.1016/j.future.2020.04.039
4. Alzahrani, F., Alharthi, S., Alotaibi, S., & Alghamdi, A. (2020). "A review of blockchain technology in cybersecurity." *Journal of Information Security and Applications*, 54, 102783.
5. Radai, Y. (1990). "The computer virus phenomenon." *Computers & Security*, 9(5), 423–430.
6. Singh, H., Kumar, N., & Singh, M. (2018). "A survey on types of malwares, propagation techniques, and their prevention." *International Journal of Advanced Research in Computer Science*, 9(5), 76–82.
7. Qadir, J., & Rajpoot, Q. M. (2019). "Anomaly detection and malware classification: A review." *Future Generation Computer Systems*, 92, 400–416. https://doi.org/10.1016/j.future.2018.10.011
8. Khandaker, R., & Khan, L. (2019). "Detection techniques for malware analysis: A comprehensive survey." *Journal of Information Security and Applications*, 49, 102377. doi: 10.1016/j.jisa.2019.102377
9. Janakiraman, R., & Jain, A. (2013). "A review of malware detection techniques." *International Journal of Advanced Research in Computer Science and Software Engineering*, 3(4), 469–474.
10. Swan, M. (2015). *Blockchain: Blueprint for a New Economy*. O'Reilly Media, Inc.
11. Buterin, V. (2014). "A next-generation smart contract and decentralized application platform." *Ethereum White Paper*, 1–36.
12. Nakamoto, S. (2008). *Bitcoin: A Peer-to-Peer Electronic Cash System.*
13. Ramzan, N., Khan, M. A., & Ahmad, A. (2017). "Blockchain-based malware detection and incident response: A case study." *International Journal of Network Security & Its Applications*, 9(6), 47–58.
14. Biswas, S., Sengupta, S., & Biswas, R. (2019). "Blockchain-based malware detection using machine learning techniques." In *2019 3rd International Conference on Inventive Systems and Control (ICISC)* (pp. 402–407). IEEE.
15. Zhang, Y., Xue, Y., Gao, Z., Yang, Y., & Liu, A. (2021). "Blockchain-enabled incident response for malware detection." *Journal of Network and Computer Applications*, 184, 103022.
16. Jang, H. J., Park, S. J., & Kim, H. J. (2020). "Blockchain-based incident response system for healthcare organizations." In *International Conference on Information and Communication Technology Convergence* (pp. 381–388). Springer.

17.  Zhang, Z., Liu, Y., Huang, W., & Rong, C. (2022). "Blockchain-enabled malware detection: A systematic literature review." *Journal of Information Security and Applications*, 67, 102999.

18.  Rafferty, L., & Miller, J. (2019). "Blockchain and incident response: Emerging opportunities and challenges." In *2019 IEEE International Conference on Blockchain (Blockchain)* (pp. 82–87). IEEE.

19.  Li, J., Yu, J., Li, X., & Chen, Y. (2020). "A blockchain-based threat intelligence sharing system for the internet of things." *Journal of Ambient Intelligence and Humanized Computing*, 11(2), 805–816.

20.  Hsiao, H., & Huang, H. (2014). "Hash-based malware detection: Techniques and challenges." *Journal of Information Science and Engineering*, 30(4), 1241–1256.

21.  Kirat, D., & Lakshmi, G. K. (2019). "YARA-based malware detection: A survey." *Journal of Network and Computer Applications*, 137, 73–91.

22.  Amini, M., & Soltanifar, M. (2017). "Malware detection based on magic number extraction and analysis." *Journal of Security and Privacy*, 1(2), 91–104.

23.  Kharraz, A., Robertson, W., & Kirda, E. (2015, May). "UNVEIL: A large-scale, automated approach to detecting ransomware." In *Proceedings of the 2015 IEEE Symposium on Security and Privacy* (pp. 1–17).

# 12 Synergizing Information Diffusion

## Exploring IoT and Blockchain Integration in Online Social Networks

*Aaquib Hussain Ganai, Nowsheena Bhat,
Rana Hashmy, Hilal Ahmad Khanday,
Mudasir Mohd, and Mohsin Altaf Wani*

Department of Computer Sciences,
University of Kashmir, J&K, India

## 12.1 INTRODUCTION

The term "social networks" was coined by Barnes in the *Human Relations Journal* in 1954 (Li, 2017). Social networks involve people and their interactions (Khatoon, 2015). The social networks of the present era have made an elite formalization. Online social networks use a dedicated website to interact with other people (Boyd, 2007). The websites made the first step towards online social networks of email, and then, with the evolution of human societies, more and more online social network platforms were created. This evolution of online social networks can be traced to Facebook and Flicker in 2004, Twitter in 2006, and Sina microblogging in 2009. Online social networks generate a huge amount of information daily, and this data serves as a base for computational and scientific work. Due to the abundance of online social network data, this data fits in the big data category (Kurha, 2015). The complexities of this data have made evolutions from fine grain to coarse grain as the web shifted from Web 1.0 to Web 2.0. From the creation of Advanced Research Projects Agency NET (ARPANET) to the creation of complex networks, whether technological (such as the Internet, world wide web, etc.) or online social networks (such as Facebook, Twitter etc.), content dissemination is the implicit means of their creation. Content dissemination is formalized as information diffusion, particularly for online social networks. The information diffusion in online social networks is a variant of an area of study in social sciences called "diffusion of innovation," which seeks to explain how, why, and at what rate new ideas and technology spread through cultures. As human beings have an innate desire to share information with others (Kumar, 2015), the wide availability of online social network services has encouraged and engaged users to share information (Kumar, 2015; Sun et al., 2019). Social networks are creating a completely virtual

DOI: 10.1201/9781003407096-12

environment that supports this challenging task of information diffusion because of their diversity and usage; 62% of adults worldwide use online social networks and spend 22% of their online time on online social networks on an average, and in India, people spend one in four minutes online using online social networking sites—more than any other Internet websites (Arnaboldi et al., 2017; Wang, 2013; Kuhnle, 2018). Thus, to mine knowledge and analyse this mined knowledge from the data of information diffusion in online social networks is the need of the hour.

Online social networks and social media analysis are popular research areas in contemporary network science, particularly when we talk of information diffusion in online social networks, whose diversity and prevalence are apparent almost everywhere—whether that can be the case of epidemics in biology, viral marketing in economics, gossip and rumour in sociology, and heat diffusion in physics, etc. (Erlandsson, 2016). Researchers try to analyse online social networks since the research discoveries by Girvan and Newman in 2002 and particularly in this subfield of information diffusion in online social networks; researchers unveiled this field of information in online social networks through varied dimensions (Cazabet, 2010). Some of the attempts related to this chapter are given here.

Guille (2013) used the T-BASIC model to study information diffusion in online social networks. Wang (2014) extended the ABXC model to the ABXCT model for online social networks. Das (2014) studied the effect of persuasion properties of messages in online social networks. Silva (2013) proposed a profile rank method for finding relevant content and influential users in online social networks. Matsubara (2012) proposed a SPIKEM model for studying the rise and fall patterns of influence propagation in online social networks. Yang (2015) used the RAIN model for modelling information diffusion in online social networks. Ren (2016) proposed a Role and Topic Aware Independent cascade model to uncover information diffusion in online social networks. Susarla et al. (2012) analysed channels central to information diffusion in online social networks. Farajtabar et al. (2015a) proposed a probabilistic model for finding out the dynamics of information diffusion and network evolution in online social networks. Dhamal et al. (2015) used independent cascade to tackle the problem of finding K users in many phases. Jiang et al. (2014) proposed a multiagent perspective for information diffusion in online social networks. Farajtabar et al. (2015b) used an associative rule learning approach for finding influential users in online social networks. Yang et al. (2018) proposed a Neural Diffusion Model of deep learning for tackling the micro-level independent cascade. Gatti et al. (2013) proposed a multiagent social network simulation for tackling information diffusion in online social networks. Kim and Yoneki (2012) proposed a hydrodynamics prediction model for information diffusion in online social networks. Davoudi and Chatterjee (2016) proposed a non-linear differential equation to study information diffusion in online social networks. Saito et al. (2013) proposed a model for hot-span detection in online social networks. Kim and Yoneki (2012) studied the influence maximization problem. P. Sermpezis et al. studied information diffusion in online social networks as epidemics. Jiang et al. (2014) proposed a game theoretic formalization for information diffusion in online social networks. Wang et al. (2013) proposed a linear diffusive model for finding the influence power of the spreading and decaying of news stories.

Obregon et al. (2019) proposed a process model based on an extension of Flexible Heuristic. Jain and Katarya (2019) used a Firefly algorithm to find global and local opinion leaders. Dai et al. (2019) proposed the Local Neighbour Communication (LNC) model to find influential users. Alemany et al. (2019) presented a new Risk model based on friend layers. For many people over the past decade, the social media platform has been their first paradigm for making new friends, discovering new material, and engaging in new types of social interactions. User's interests, connections, behaviours, content, and location can all be mined for valuable information that can be used for marketing and other purposes. The social media sites open the door for targeted advertising. The most popular social media platforms nowadays are aggregators allowing users to monetize their material effectively. Privacy and data security are significant issues for online communities. Blockchain's decentralized ledger and philosophy encrypt user data to keep it safe. Social networking services, including Whatsapp, iMessage, Signal, Wire, Threema, etc., use end-to-end encryption. The problem arises when users communicate metadata and communications, allowing hackers to acquire private information. Social media platforms built on the blockchain provide more than increased privacy and security. Blockchain makes it easier for users to have editorial authority over their data. Blockchain technology in social networks is a decentralized method for privacy protection, e-commerce, crowdfunding, and intelligent apps and contracts. Nowadays, it's easy to get online and connect with others for free due to the proliferation of social networking sites like Facebook, LinkedIn, Instagram, Twitter, Reddit, etc. In the social network company era, advertising and analytics are where the money lies, despite the apparent importance of development and server infrastructures to social networking platforms. Fake news (Facebook), trolling (Twitter), and censorship and demonetization (YouTube) are some of the most fundamental issues with social networking. Some networks and the associated aspects can only be studied using models based on blockchain technology (Guidi and Michienzi, 2021). The blockchain models can handle the sentimental perspective of information diffusion in online social networks. The information diffusion studies in social network analysis (SNA) have attempted to use the current blockchain technological concepts. The researchers have modelled the handling of information diffusion in online social networks using blockchain technology-augmented models.

Some of the attempts using the blockchain models are as follows: Wu et al. (2021) studied the community detection in blockchain social networks using the modified clustering algorithm. Thakur and Breslin (2021) proposed the decentralized method that is based on bitcoin blockchain model to prevent social bolts from spreading rumours. Liu and Bin (2022) proposed the influence maximization model in blockchain-based social networks that are based on linear threshold model.

The information diffusion in online social networks can be handled using IoT-based models or blockchain-based models or by integrating both. The use of social media is deeply embedded in daily life. While user data is stored, most popular social networks have a centralized system paradigm that raises serious issues such as content ownership and excessive commercialization. Blockchain-based decentralized social networks act as an alternative to address these issues. The decentralized social networks are gaining traction now as their underlying structures can compete effectively with those of centralized options. Traditional features like posting and commenting and more innovative features like voting and rewards systems can all be obtained via decentralized social networks. Staking systems based on cryptocurrency tokens enable a robust ecosystem for influencers to engage with their audiences. Today's introduction of Web 3.0 led to the development of decentralized autonomous management and blockchain-based models. In particular, technological development has further spawned a distinct type of social platform called blockchain-enabled social media (i.e., SocialFi), which is expanding in size and user base. Blockchain is gaining popularity as a trusted database format for various applications. The Internet of Things (IoT), social media, robots, government, supply chain, and even healthcare organizations use blockchain. Streaming social media data presents new challenges for secure, private, and trustworthy storage and analysis as blockchain research grows in popularity. In this chapter, we explore the phenomena of information diffusion in social networking with the IoT and blockchain.

## 12.2   INFORMATION DIFFUSION IN ONLINE SOCIAL NETWORKS

Information diffusion in online social networks revolves around four components— (1) people; (2) content; (3) network; and (4) diffusion depicted in Table 12.1.

Each component acts as a full-fledged dimension for researchers of information diffusion in online social networks to tackle the problems intrinsic to these components. People are the fundamental physical entities that are apparent on online social networks. Their reflections are like user accounts, pages, handles, etc. The graphs represent online social networks, and people act modelled as graph

**TABLE 12.1**
**Components of Information Diffusion in Online Social Networks (OSNs)**

| Component | Representation | Function |
|---|---|---|
| People | | These are the actors of scene who produce information (content) for diffusion in OSNs. |
| Content | | This is the element that people create for diffusions in OSNs. |
| Network | | This is the output that people with diffusing contents create in OSNs. |
| Information Diffusion | | This is process that can reveal varied pattern formations that have led to network formations in OSNs. |

nodes. Influential user detection in online social networks is the most widespread problem studied in information diffusion under this domain. Much research has been done on this problem area of information diffusion in online social networks. The online social networks are modelled as graphs, with adopted weights as communication contents; therefore, online social networks can be represented by a graph G(V, E) where V is the set of nodes as people (users) and E is the set of edges representing communication (Khatoon and Banu, 2015). These online social graphs can be directed or undirected based on the characteristics of underlying online social network. These online social graphs can be represented by an adjacency list, adjacency matrix, incidence lists, incidence matrix; use list representation when the complexity of online social network under the scanner is slight and use matrices when the network is dense.

When a piece of content (information) flows from one individual to another or from one community to another in an online social network, an act of information diffusion is said to have occurred in that online social network (Li et al., 2017). When this act of information diffusion in online social networks is studied under the broad domain of information diffusion of SNA, we often encounter the "3 W issue" as depicted in Table 12.2; the three W's are "what," "why," and "where." The first w, "what," refers to the question: "what information is there to be found in online social networks?" The second w is "why," which refers to the question "why has the information propagated this way?" The third w is "where," which refers to the question: "where will the information be diffused in the future?" (Li et al. 2017). This three W issue is best studied under the broad spectrum of two information diffusion process models: explanatory models and prediction models.

### 12.2.1 EXPLANATORY MODELS

Such models explain the information diffusion process adopted by the given online social network under the scanner. These models reflect real human behaviour. The best example of these models is epidemic models such as SIS, SIR, etc. (Cole, 2011). Parameters of these models are easily available and are a little complex in case of modelling complexities (Arnaboldi et al., 2017).

### 12.2.2 PREDICTION MODELS

Such models long to tackle the decision problem, which can take "yes" or "no" states for deciding whether diffusion will usher or not from the current known state to the unknown state; the unknown state belongs to the future. These models do not reflect the true nature of human socialism. Parameters of models are hard to model, so these models are highly complex in nature. The best examples of these models are the independent cascade model, linear threshold model, etc(Arnaboldi et al., 2017).

Researchers have used both models to put their efforts into information diffusion in online social networks, keeping the two model paradigms as a base. Table 12.2 reflects the actual functionalities of the two-process model on information diffusion in online social networks.

**TABLE 12.2**
**Three-W Issue and the Related Capabilities of Modelling the Human Behaviour**

| Model of Information Diffusion | Tackle W1 of 3W Issue | Tackle W2 of 3W Issue | Tackle W3 of 3W Issue | Reflect True Human Behaviour |
| --- | --- | --- | --- | --- |
| Explanatory model | Yes | Yes | No | Yes |
| Prediction model | Yes | Yes | Yes | No |

## 12.3   INTERNET OF THINGS IN SOCIAL NETWORKING

The IoT completely transforms how we communicate and interact in online social networks. IoT builds a harmonious blend of digital and physical experiences by incorporating IoT devices and sensors into our daily routines (Reshi and Sholla, 2022). Increased transmission demands on the web are a direct result of the proliferation of IoT intelligent terminals and applications made possible by 5G communication technology. It is crucial to choose trustworthy relay users to guarantee data transmission security in opportunistic social networks, which use the "store-carry-forward" mechanism to finalize message delivery. Through the use of real-time updates, users of social networking sites can facilitate more relevant and contextual interactions with one another. Improved event attendance, location-based networking, and even the formation of online communities based on shared interests enabled by the IoT are all possible outcomes of this development.

However, there are several drawbacks of incorporating the IoT into social networking. The increased threat to users' personal information and privacy is a significant issue. Large volumes of personal data generated by IoT devices can be compromised and used maliciously without adequate security measures. Identity theft, unauthorized surveillance, and invasive data collection are some problems that might result from the constant flow of data between devices and social networks, which presents possible entry points for evil activities to exploit.

## 12.4   BLOCKCHAIN IN SOCIAL NETWORKING

The advent of blockchain technology offers hope for addressing the difficulties brought about by the IoT in social networking interactions. Blockchain is an immutable digital record that can maintain data trustworthiness, safety, and accuracy. Social media sites may create an immutable record of all data created by IoT devices; thus, users could exercise greater discretion over the sharing and use of their data. Blockchain's distributed ledger structure also makes it harder for hackers to compromise the system by eliminating a single point of failure. Furthermore, smart contracts built on the blockchain can enable private and automated data-sharing arrangements between IoT gadgets and social media sites. In this way, users can rest assured that their privacy preferences are considered across all interactions with the service. In addition, the cryptographic concepts behind blockchain can improve authentication and authorization procedures, reducing the likelihood that IoT devices and the private data they collect will be compromised.

While the IoT has the potential to enhance social networking connections, it also has drawbacks that must be addressed, such as privacy and security problems. Blockchain technology provides a firm answer by creating a trustworthy and open network where users can continue to have discretion over their data and communications. By implementing blockchain technology, social media networks may protect their users' data and privacy without sacrificing the advantages of IoT integration.

There is great potential for increased social network security with blockchain technology and the IoT. The expansion of IoT devices and associated data exchanges pose significant security challenges, but blockchain provides a decentralized and tamper-proof ledger to address these issues. Blockchain's use of cryptography and consensus procedures guarantees the truthfulness and accuracy of data produced by IoT devices, laying a solid groundwork for safe communications on social media.

Due to its distributed structure, blockchain has no central point of failure like other centralized systems. In the context of social media, this means that hackers cannot target a specific entry point to steal information or invade users' privacy. Because the ledger's transactions are stored over a distributed network of computers, tampering with the data is extremely difficult. Improved security makes IoT-driven social interactions less vulnerable to attacks like hacking, data breaches, and fraud.

The capacity of blockchain to create and enforce transparent and self-executing smart contracts helps to improve security as well. These smart contracts can govern the circumstances under which data is shared between IoT devices and social media platforms. This function is invaluable in protecting users' privacy because it gives them more say over how their data is used and shared in the social networking ecosystem.

In addition, the immutable record of transactions that blockchain provides serves as an additional safeguard. Participants in the social networking environment have access to a verifiable form of all interactions, including the ability to track the data's origin and its journey across the network. The capacity to quickly determine the point of compromise and take corrective action in the event of a security issue is made possible by this traceability.

## 12.5   INTEGRATION OF IOT AND BLOCKCHAIN IN SOCIAL NETWORKING

Online social networks are created for social interactions. Since the emergence of online social networks, online social interactions have evolved from simple talkie messages to audiovisual messages, making social interactions very diverse. As these interactions became diverse, the information carried also flooded the online social networks, thereby creating diffusions of this information which, internally, made numerous phenomena among the users of online social networks. These phenomena are so diverse that even a tiny change in the scale of the view can flip phenomena. This diversity of one-to-one correspondence from views to adopted phenomena has made researchers take a different perspective on information diffusion in online social networks by modelling information cascades, finding popular content, tackling the problem of finding influential users, studying the effects of external factors on the diffusion of information in online social network and semantic analysis according to the view of blockchain technology. Combining blockchain and IoT creates a decentralized, tamper-proof, and transparent framework that significantly improves social networking security. Data generated by IoT devices may now be shared securely and privately due to the blockchain's cryptographic algorithms, consensus processes, smart contracts, and audit trail features. This creates a safer and more reliable setting for social networking interactions, giving users more outstanding agency in an increasingly complex digital world.

Blockchain is gaining popularity as a trusted database format for various applications. The IoT, social media, robots, government, the supply chain, and even healthcare organizations use blockchain. Streaming social media data presents new challenges for secure, private, and trustworthy storage and analysis as blockchain research grows in popularity. This research adapts the streaming social media data set for use with blockchain systems and analyses the resulting data within the context of this issue. A novel technique called the Dense Region Detection Algorithm (DRDA) technique has been developed to identify highly populated areas in cities. To determine the dense regions of New York City, miners take streaming data from the city and run the proposed DRDA (Ecemis, 2022). Zhao (2022) considers the standard social network propagation model to the question of how blockchain technology may affect social media. Social network factions are categorized as "support nodes" or "opposition nodes." It employs the evolutionary game theory to determine the likelihood of transition between states and examine the effect of population density on the rate of information spread throughout a blockchain network. The impact of state transition probability on network group density and time to steady state is then analysed. Finally, it investigates whether or not a blockchain-based social network's incentive strategy plays a role in its users' spreading behaviour. Traditional features like posting and commenting and more innovative features like voting and rewards systems can all be obtained via decentralized social networks. Staking systems based on cryptocurrency tokens enable a robust ecosystem for influencers to engage with their audiences. Decentralized social networks contain a wealth of information that can be used to understand human behaviour in several novel ways. However, it is not simple to access and collect data from these social networks due to the need for in-depth knowledge of blockchain technology, which is not the primary focus of researchers in computer science and social science. Nguyen (2022) proposed the SoChainDB architecture to make data acquisition from emerging social media platforms more accessible. One of the largest blockchain-based social networks, Hive, is crawled, and its data is published to demonstrate the power and capabilities of SoChainDB. They perform in-depth analysis to learn about the hidden meanings in Hive data and cover some exciting uses of Hive, like games and a market for non-fungible tokens. This architecture can quickly adapt to blockchain-based social networks.

As a result, these blockchain-enabled social media companies' quick rise highlights the need to comprehend better the causes of this growth and the cutting-edge tactics and business strategies companies use in this new industry (Zhan, 2023). Jia (2023) proposed a new online social network data mining model based on blockchain to address the issues of low security and accuracy of traditional online social network data mining models. The hash function in the blockchain encrypts the data from online social networks based on the blockchain's structure to increase data security. The fuzzy covariance matrix derives the data clustering centre from the encrypted data to complete the data clustering. The time series analysis method is used to mine data from online social networks by the results of data clustering. According to the results, this model significantly increases data security from online social networks and accuracy. The authors analysed the various forms of private information shared on social media and designed data protection schemes for multiple forms of privacy leakage by decentralization, non-tampering, hash function, asymmetric cryptographic signature algorithm, and other features of blockchain technology. The privacy protection scheme based on shared data effectively anonymizes the input and an output user, encrypts the signature, and verifies the shared information. The privacy protection scheme based on primary personal data can effectively anonymize user information and isolate the connection between user information and individual users (Wang. 2023).

## 12.6   CONCLUSION

Integrating blockchain and IoT is the attractive strategy for bolstering online safety, especially in social media. This integration uses the immutable and distributed ledger of blockchain to solve serious problems with data integrity and privacy that arise from the widespread use of IoT devices and the sharing of personal information in social media. Through cryptographic algorithms, consensus processes, and self-executing smart contracts, blockchain provides data authenticity, integrity, and regulated sharing. This integration improves safety, lessens privacy worries, and builds a solid platform for online communication.

## REFERENCES

Alemany, Jose, Elena Del Val, Juan M. Alberola, and Ana Garćia-Fornes. "Metrics for privacy assessment when sharing information in online social networks." *IEEE Access* 7 (2019): 143631–143645.

Arnaboldi, Valerio, Marco Conti, Andrea Passarella, and Robin I. M. Dunbar. "Online social networks and information diffusion: The role of ego networks." *Online Social Networks and Media* 1 (2017): 44–55.

Boyd, Danah M., and Nicole B. Ellison. "Social network sites: Definition, history, and scholarship." *Journal of computer-mediated Communication* 13, no. 1 (2007): 210–230.

Cazabet, Remy, Frederic Amblard, and Chihab Hanachi. "Detection of overlapping communities in dynamical social networks." In *2010 IEEE Second International Conference on Social Computing*, pp. 309–314. IEEE, 2010.

Cole, William David. *An Information Diffusion Approach for Detecting Emotional Contagion in Online Social Networks*. Arizona State University, 2011.

Dai, Jinying, Bin Wang, Jinfang Sheng, Zejun Sun, Faiza Riaz Khawaja, Aman Ullah, Dawit Aklilu Dejene, and Guihua Duan. "Identifying influential nodes in complex networks based on local neighbor contribution." *IEEE Access* 7 (2019): 131719–131731.

Das, Abhimanyu, Sreenivas Gollapudi, and Emre Kiciman. *Effect of Persuasion on Information Diffusion in Social Networks.* Retrieved from research. microsoft. com/pubs/217325/persuasion_2014-05-19. pdf (2014).

Davoudi, Anahita, and Mainak Chatterjee. "Prediction of information diffusion in social networks using dynamic carrying capacity." In *2016 IEEE International Conference on Big Data (Big Data)*, pp. 2466–2469. IEEE, 2016.

de C Gatti, Maira A., Ana Paula Appel, Cicero Nogueira dos Santos, Claudio Santos Pinhanez, Paulo Rodrigo Cavalin, and Samuel Barbosa Neto. "A simulation-based approach to analyze the information diffusion in microblogging online social network." In *2013 Winter Simulations Conference (WSC)*, pp. 1685–1696. IEEE, 2013.

Dhamal, Swapnil, and Y. Narahari. "A multi-phase approach for improving information diffusion in social networks." *arXiv preprint arXiv:1502.06133* (2015).

Ecemis, Alper, Ahmet Sakir Dokuz, and Mete Celik. "Dense region detection of streaming social media datasets using blockchain-based secure computation." In *2022 3rd International Informatics and Software Engineering Conference (IISEC)*, pp. 1–5. IEEE, 2022.

Erlandsson, Fredrik, Piotr Bródka, Anton Borg, and Henric Johnson. "Finding influential users in social media using association rule learning." *Entropy* 18, no. 5 (2016): 164.

Farajtabar, Mehrdad, Yichen Wang, Manuel Gomez Rodriguez, Shuang Li, Hongyuan Zha, and Le Song. "Coevolve: A joint point process model for information diffusion and network co-evolution." *Advances in Neural Information Processing Systems* 28 (2015a).

Farajtabar, Mehrdad, Manuel Gomez-Rodriguez, Yichen Wang, Shuang Li, Hongyuan Zha, and Le Song. "Co-evolutionary dynamics of information diffusion and network structure." In *Proceedings of the 24th International Conference on World Wide Web*, pp. 619–620. 2015b.

Guidi, Barbara, and Andrea Michienzi. "The decentralization of social media through the blockchain technology." In *Companion Publication of the 13th ACM Web Science Conference 2021*, pp. 138–139. 2021.

Guille, Adrien, Hakim Hacid, Cecile Favre, and Djamel A. Zighed. "Information diffusion in online social networks: A survey." *ACM Sigmod Record* 42, no. 2 (2013): 17–28.

Han, Jiawei, Micheline Kamber, and P. E. I. Jian. *Data Mining Concepts and Techniques, File.* C:/Users/Pascale/AppData/Local/Microsoft/Windows/INetCache/IE/BBV966TF/The-Morgan-Kaufmann-Series-in-Data-Management-Systems-Jiawei-Han-Micheline-Kamber-Jian-Pei-Data-Mining.-Concepts-and-Techniques-3rd-Edition-Morgan-Kaufmann-2011. pdf (2012).

Jain, Lokesh, and Rahul Katarya. "Discover opinion leader in online social network using firefly algorithm." *Expert Systems with Applications* 122 (2019): 1–15.

Jia, Xiaoqiang. "Construction of online social network data mining model based on blockchain." *Soft Computing* 27, no. 8 (2023): 5137–5145.

Jiang, Chunxiao, Yan Chen, and K. J. Ray Liu. "Modeling information diffusion dynamics over social networks." In *2014 IEEE International Conference on Acoustics, Speech and Signal Processing (ICASSP)*, pp. 1095–1099. IEEE, 2014.

Khatoon, Mehjabin, and W. Aisha Banu. "A survey on community detection methods in social networks." *International Journal of Education and Management Engineering* 5, no. 1 (2015): 8.

Kim, Hyoungshick, and Eiko Yoneki. "Influential neighbours selection for information diffusion in online social networks." In *2012 21st International Conference on Computer Communications and Networks (ICCCN)*, pp. 1–7. IEEE, 2012.

Kuhnle, Alan, Md Abdul Alim, Xiang Li, Huiling Zhang, and My T. Thai. "Multiplex influence maximization in online social networks with heterogeneous diffusion models." *IEEE Transactions on Computational Social Systems* 5, no. 2 (2018): 418–429.

Kumar, Kp Krishna. *Information Diffusion Modelling to Counter Semantic Attacks in Online Social Networks.* PhD diss., 2015.

Kurka, David Burth. *Online Social Networks: Knowledge Extraction from Information Diffusion and Analysis of Spatio-Temporal Phenomena Redes Sociais Online: Extraçao de conhecimento e análise espaço-temporal de.* PhD diss., [sn], 2015.

Li, Mei, Xiang Wang, Kai Gao, and Shanshan Zhang. "A survey on information diffusion in online social networks: Models and methods." *Information* 8, no. 4 (2017): 118.

Liu, Chang, and Sheng Bin. "Research on maximizing influence of blockchain social network based on BCLT model." *Discrete Dynamics in Nature and Society* 2022 (2022).

Matsubara, Yasuko, Yasushi Sakurai, B. Aditya Prakash, Lei Li, and Christos Faloutsos. "Rise and fall patterns of information diffusion: Model and implications." In *Proceedings of the 18th ACM SIGKDD International Conference on Knowledge Discovery and Data Mining*, pp. 6–14. 2012.

Nguyen, Hoang H., Dmytro Bozhkov, Zahra Ahmadi, Nhat-Minh Nguyen, and Thanh-Nam Doan. "SoChainDB: A database for storing and retrieving blockchain-powered social network data." In *Proceedings of the 45th International ACM SIGIR Conference on Research and Development in Information Retrieval*, pp. 3036–3045. 2022.

Obregon, Josue, Minseok Song, and Jae-Yoon Jung. "InfoFlow: Mining information flow based on user community in social networking services." *IEEE Access* 7 (2019): 48024–48036.

Ren, Xiaoxuan, and Yan Zhang. "Predicting information diffusion in social networks with users' social roles and topic interests." In *Information Retrieval Technology: 12th Asia Information Retrieval Societies Conference, AIRS 2016, Beijing, China, November 30–December 2, 2016, Proceedings 12*, pp. 349–355. Springer International Publishing, 2016.

Reshi, Iraq Ahmad, and Sahil Sholla. "Challenges for security in iot, emerging solutions, and research directions." *International Journal of Computing and Digital Systems* 12, no. 1 (2022): 1231–1241.

Saito, Kazumi, Masahiro Kimura, Kouzou Ohara, and Hiroshi Motoda. "Detecting changes in information diffusion patterns over social networks." *ACM Transactions on Intelligent Systems and Technology (TIST)* 4, no. 3 (2013): 1–23.

Silva, Arlei, Sara Guimarães, Wagner Meira Jr, and Mohammed Zaki. "ProfileRank: Finding relevant content and influential users based on information diffusion." In *Proceedings of the 7th Workshop on Social Network Mining and Analysis*, pp. 1–9. 2013.

Sun, Qingsong, Ying Li, Haibo Hu, and Shulin Cheng. "A model for competing information diffusion in social networks." *IEEE Access* 7 (2019): 67916–67922.

Susarla, Anjana, Jeong-Ha Oh, and Yong Tan. "Social networks and the diffusion of user-generated content: Evidence from YouTube." *Information Systems Research* 23, no. 1 (2012): 23–41.

Thakur, Subhasis, and John G. Breslin. "Rumour prevention in social networks with layer 2 blockchains." *Social Network Analysis and Mining* 11 (2021): 1–17.

Wang, Chengjun. *Jumping Over the Network Threshold: Information Diffusion on Information Sharing Websites*, PhD diss., City University of Hong Kong, 2014.

Wang, Feng, Haiyan Wang, Kuai Xu, Jianhong Wu, and Xiaohua Jia. "Characterizing information diffusion in online social networks with linear diffusive model." In *2013 IEEE 33rd International Conference on Distributed Computing Systems*, pp. 307–316. IEEE, 2013.

Wang, Pingshui, Jianwen Zhu, and Qinjuan Ma. "Privacy data protection in social networks based on blockchain." (2023).

Wu, Sissi Xiaoxiao, Zixian Wu, Shihui Chen, Gangqiang Li, and Shengli Zhang. "Community detection in blockchain social networks." *Journal of Communications and Information Networks* 6, no. 1 (2021): 59–71.

Yang, Cheng, Maosong Sun, Haoran Liu, Shiyi Han, Zhiyuan Liu, and Huanbo Luan. "Neural diffusion model for microscopic cascade prediction." *arXiv preprint arXiv:1812.08933* (2018).

Yang, Yang, Jie Tang, Cane Leung, Yizhou Sun, Qicong Chen, Juanzi Li, and Qiang Yang. "Rain: Social role-aware information diffusion." In *Proceedings of the AAAI Conference on Artificial Intelligence* 29, no. 1. 2015.

Zhan, Yuanzhu, Yu Xiong, and Xinjie Xing. "A conceptual model and case study of blockchain-enabled social media platform." *Technovation* 119 (2023): 102610.

Zhao, Yan, Sheng Bin, and Gengxin Sun. "Research on information propagation model in social network based on BlockChain." *Discrete Dynamics in Nature and Society* 2022 (2022).

# Index

For Product Safety Concerns and Information please contact our EU
representative GPSR@taylorandfrancis.com
Taylor & Francis Verlag GmbH, Kaufingerstraße 24, 80331 München, Germany

www.ingramcontent.com/pod-product-compliance
Lightning Source LLC
Chambersburg PA
CBHW060831170526
45158CB00001B/135